AF678797

The Problem of Soil Contamination: Causes, Effects, and Solutions from Environmental Science

THE PROBLEM OF SOIL CONTAMINATION: CAUSES, EFFECTS, AND SOLUTIONS FROM ENVIRONMENTAL SCIENCE

Vierah Hulley

Delve PUBLISHING

www.delvepublishing.com

The Problem of Soil Contamination: Causes, Effects, and Solutions from Environmental Science

Vierah Hulley

Delve Publishing
224 Shoreacres Road
Burlington, ON L7L 2H2
Canada
www.delvepublishing.com
Email: orders@arclereducation.com

© 2023 Delve Publishing

ISBN: 978-1-77469-455-8 (Hardcover)

This book contains information obtained from highly regarded resources. Reprinted material sources are indicated and copyright remains with the original owners. Copyright for images and other graphics remains with the original owners as indicated. A Wide variety of references are listed. Reasonable efforts have been made to publish reliable data. Authors or Editors or Publishers are not responsible for the accuracy of the information in the published chapters or consequences of their use. The publisher assumes no responsibility for any damage or grievance to the persons or property arising out of the use of any materials, instructions, methods or thoughts in the book. The authors or editors and the publisher have attempted to trace the copyright holders of all material reproduced in this publication and apologize to copyright holders if permission has not been obtained. If any copyright holder has not been acknowledged, please write to us so we may rectify.

Notice: Registered trademark of products or corporate names are used only for explanation and identification without intent of infringement.

Delve Publishing publishes wide variety of books and eBooks. For more information about Delve Publishing and its products, visit our website at www.delvepublishing.com.

ABOUT THE AUTHOR

Vierah Hulley (PhD) is an international expert in Environmental and Earth Sciences, with an extensive vocational background in the areas of environmental policy development, environmental and social risk management, natural resources management, sustainability, and climate risks management. She is the Founder of HL Nexus: Sustainable Management Solutions; a company dedicated to harnessing the power of data science in managing sustainability and climate risks. Vierah holds a master's degree in Geology and a PhD in Hydrogeology, with a focus on environmental and spatial science.

TABLE OF CONTENTS

List of Figures .... *xi*

Abstract .... *xvii*

Preface .... *xix*

Chapter 1 Soil and its Role in the Ecosystem **1**

1.1. Introduction to Soil as an Ecosystem Service 2

1.2 Medium for Plant Growth 4

1.3 Regulating Carbon and Nitrogen Cycles 6

1.4 Soils and Climate Change 8

1.5 Regulating Water Supplies 9

1.6 Recyclers of Raw Materials 12

1.7 Habitat for Soil Organisms 13

1.8 Engineering and Landscape Medium 15

1.9 Cultural and Physical Heritage 16

Chapter 2 Soil Pollution **19**

2.1 Soil Contamination Vs. Soil Pollution 20

2.2 Man-Made Pollutants 22

2.3 Common Soil Contaminants of Concern and their Fate in The Soil Environment 26

2.4 Processes that Cause Soil Contamination 30

Chapter 3 Impacts of Soil Contamination. **33**

Introduction 34

3.1 Human Health Impacts 34

3.2. Ecotoxicological Impacts on Terrestrial Ecosystems 36

3.3. Impacts on Aquatic Ecosystems 37

3.4. Food Chain Disruptions 37

3.5. Air Quality Impacts 37
3.6. Groundwater Contamination 38
3.7. Socio-Economic Impacts 38
3.8. Soil Pollution: Possible Solutions 39
3.9 Causes of Soil Pollution 40
3.10 Industrial Accidents 57
3.11 Intensive Farming 59
3.12 Petrochemicals 62
3.13 Road Debris 65
3.14 Drainage of Contaminated Surface Water Into the Soil 68
3.15 Waste Disposal 71

Chapter 4 Consequences of Soil Pollution 77
4.1 What Is Soil Pollution? 78
4.2 Soil Pollution Causes and Types 78
4.3 Soil Pollution Consequences 80
4.4 Significant Impacts of Soil Pollution According to IPBES and the Fao 80
4.5 Damage to Health 81
4.6 Poor Harvest 87
4.7 Climate Change 90
4.8 Water and Air Pollution 93
4.9 Population Displacement 96
4.10 Soil Contamination and Species Extinction 99
4.11 Biodiversity Conservation and Poverty 101
4.12 Economic Impact 102

Chapter 5 How Contaminants Affect Soil Functions 107
5.1 Interaction With the Environment 115
5.2 Mapping Soil Functions 117
5.3 Categories of Contaminated Lands 119
5.4 Microplastics 121
5.5 Hydrocarbons 124
5.6 Electronic Waste 128
5.7 Sites of Special Concern 131
5.8 Exposure to Contaminant 133

5.9 Pesticides and Herbicides 134
5.10 Conclusion 137

Chapter 6 How are People Exposed to Soil Contaminants? 139
6.1 Ingesting Dust or Soil Directly 140
6.2 Ingesting Food or Vegetables Grown in Contaminated Soil or with Foods in Contact with Contaminants 146
6.3 Skin Contact with Dust or Soil 155
6.4 Vapors from the Soil 158
6.5 Inhaling Dusts While Working in Soil or Windy Environments 162
6.6 Exposure to Contaminated Ground Water 164

Chapter 7 Soil Contamination and Sites of Special Concern 167
7.1 Industrial And Manufacturing Sites 168
7.2 Industrial Activities 170
7.3 Manufacturing and Recycling 172
7.4 Transport 173
7.5 Workplace Accidents 174
7.6 Landfills, Junkyards and Waste Disposal Sites 176
7.7 Types of Landfills 178
7.8 A Specially Designed Landfill 180
7.9 Highway Corridors, Parking Lots, Areas of Heavy Traffic 180
7.10 Pollution 181
7.11 The Most Effective Gardening Method 190
7.12 Former Farmland with the Build-Up of Contaminants 191

Chapter 8 Most Polluted Soil in the World 195
8.1 Introduction 196
8.2 Soil Pollution 197
8.3 Agricultural Areas 200
8.4 Polluted Soil in Industrial Areas 204
8.5 Most Polluted Soil By Radioactive Elements 208
8.6 Conclusion 213

Chapter 9 **Risk Assessments and the Remediation of Contaminated Soils 215**

9.1 Introduction 216

9.2 Bioremediation 217

9.3 Long-Term Development and Restorative Land Use in Underprivileged Regions 220

9.4 The Size and Environmental Impact of Bioremediation Soil Pollution in Disadvantaged Communities 221

9.5 How Soil Characteristics Influence Biodegradation 222

9.6 Environmental and Human Health Risks of Contaminated Soils 224

9.7 Toxaphene 224

9.8 Toxic Heavy Metals 226

9.9 Bio-Stimulant 229

9.10 Whey 232

9.11 Bioaugmentation 235

9.12 Eco-Friendly Technologies 237

9.13 Technology 241

9.14 Environmental Science and Technology 242

9.15 Conclusions 243

Chapter 10 **Soil Protection Policy 245**

10.1 What Is Soil Protection? 246

10.2 European Union (EU) Soil Policy 248

10.3 Existing Laws on Soil Protection 250

10.4 The European Green Deal 252

10.5 The Resource Conservation and Recovery Act (RCRA) 254

10.6 Assessment of Current Soil Protection Strategies 255

10.7 RCA Reports Usefulness Can Be Improved 256

10.8 Longstanding Social Perspectives and Customs 257

References 259

Index 265

LIST OF FIGURES

Figure 1.1: Soil is a vital resource in the global ecosystem
Figure 1.2: Soil contains various components
Figure 1.3: Soil anchors crops to the ground
Figure 1.4: Soil helps maintain a balance of carbon and nitrogen in the atmosphere
Figure 1.5: Placing land under cultivation contributes to the greenhouse effect
Figure 1.6: Soil enables the preservation of water for future use
Figure 1.7: Decomposed materials are used by plants to bring forth new life
Figure 1.8: There are various kinds of organisms that live in soil
Figure 1.9: Soil is used in building various infrastructure
Figure 1.10: Soil is a cover for the earth
Figure 2.1: In Senegal, a pastoralist herds his cattle past a mound of garbage.
Figure 2.2: Changed Soil properties due to pollution
Figure 2.3: Man made soil pollution
Figure 2.4: Pollutants that Contaminate Soil
Figure 2.5: Toxic Metals that Cause Soil Pollution
Figure 2.6: Cigarette Butt
Figure 2.7: Herbicides application
Figure 2.8: Different Sources of pollution.
Figure 3.1: Sample elements of micro-plastics
Figure 3.2: The environmental fate and impact of microplastics on soil resulting in a larger effect on the ecosystem
Figure 3.3: Using magnets is one way of removing microplastics while mitigating their effect on soil
Figure 3.4: Oil spills are among modern man's problem in the care of soils
Figure 3.5: Oil spill on the land has a higher effect on soil.
Figure 3.6: Effluents from acid mine drainage: example of Rio Tinto, Spain
Figure 3.7: The adverse effects of accidental spills
Figure 3.8: A corroded tank in a mine
Figure 3.9: The adverse effects of acid rain on soil

Figure 3.10: An illustration of the effects of agrochemicals on soil

Figure 3.11: Effects of artillery and contamination

Figure 3.12: A piece of land bearing the adverse effects of intensive farming

Figure 3.13: An overview of petrochemicals

Figure 3.14: A sample image of road debris

Figure 3.15: Contaminated surface water can run and eventually affect the soil

Figure 4.1: Soil Pollution

Figure 4.2: Ecosystems impairment caused by soil pollution

Figure 4.3: What Are The Harmful Effects Of Soil Pollution?

Figure 4.4: Health effects of pollution

Figure 4.5: Soil pollution and risk to human health

Figure 4.6: (**a**) Aerial view of a typical erosion gully in the Cerrado of Brazil. On the right, soybean plantation showing bare soils after harvest, on the left, riparian vegetation surrounding a stream. In the foreground, an earth road and two erosion gullies of different ages. The erosion gullies breach the riparian vegetation (capture date of photographs July 1995); (**b**) Active erosion gully in Figure (**a**). Deep lateral piping hollows drain groundwater from the surroundings, which flows off through the channel even during the dry season; (**c**) Stream receiving sediment from the active erosion gully in Figure (**b**) during the dry season.; (**d**) Pioneer vegetation developing on the bottom of the erosion gully in **Figure 1c** during the dry season. © Karl M. Wantzen.

Figure 4.7: Soil Erosion – Causes and Effects

Figure 4.8: Land degradation worsening climate change- serious land degradation in Nauru after the depletion of the phosphate cover through mining

Figure 4.9: China's Fight against Climate Change and Environmental Degradation

Figure 4.10: Water Pollution and Soil Pollution

Figure 4.11: People's displacement: Villagers fleeing gunfire in a camp for internally displaced persons during the 2008 Nord-Kivu war

Figure 4.12: Sixth mass extinction of wildlife accelerating

Figure 4.13: Combat Desertification

Figure 4.14: Socio-economic impacts of soil pollution

Figure 5.1: Fundamental soil functions in ecosystems

Figure 5.2: Correspondence between soil functions and ecosystem services

Figure 5.3: Urban Development

Figure 5.4: Functions of the Soil

Figure 5.5: Human-Environment Interactions and Land Use Dynamics

Figure 5.6: Palmyra

Figure 5.7: An example of soil mapping

Figure 5.8: (a) Allocation of additional observations (AO) to field observations (FO) in Google Earth for the soil mapping unit (SMU) TypicEndoaquent+FluventicDystrudept. (b) Original and final proportions of the training datasets by SMU.

Figure 5.9: Effects of microplastics in the soil environment

Figure 5.10: Development and application of tests for microplastic detection in soil

Figure 5.11: Despite the fact that the majority of plastic pollution originates on land, the current study is mostly focused on aquatic habitats.

Figure 5.12: This photo shows oil floating on water of contaminated soils south of New Orleans

Figure 5.13: Soil contaminated by an oil spill: an example off the coast of Mauritius (Indian Ocean)

Figure 5.14: Kelp after an oil spill

Figure 5.15: The health risk associated with contaminated soil will depend on the type of hazardous substance, the exposure route and the length of time a person is exposed.

Figure 6.1: Children are most likely to ingest soil as they play

Figure 6.2: Food supplies such as beans and carrots can be contaminated when exposed to contaminated soils

Figure 6.3: Soil pica is the ingestion of soil

Figure 6.4: Children's hands are likely to carry lots of contaminated soil

Figure 6.5: Continuous consumption of contaminated soils leads to build dup of contaminants in the body

Figure 6.6: Vegetables growing in contaminated areas are likely to absorb some of the contaminants

Figure 6.7: Vegetable contamination can occur through the supply of contaminated water into soil

Figure 6.8: Presence of large amounts of minerals in foods is linked to the presence of contaminants

Figure 6.9: Some vegetables in the market contain contaminants from the soil

Figure 6.10: Dermal exposure is exposure to soil contaminants through the skin

Figure 6.11: Dust particles settled on skin surfaces may be absorbed into the body

Figure 6.12: Vapors from soil are mostly released through evaporation

Figure 6.13: Exposure to soil contaminants in workplaces is known as an occupational hazard

Figure 6.14: Groundwater contaminated with soil particles is hazardous to human health

Figure 7.1: Industrial manufacturing facility plant modern factory

Figure 7.2: Mining, Quarrying, and Oil and Gas Extraction

Figure 7.3: Nickel

Figure 7.4: Common soil contamination sources

Figure 7.5: Landfills

Figure 7.6: Invasive species in rubbish dumps

Figure 7.7: Areas of heavy traffic

Figure 7.8: Oil spill on a car

Figure 7.9: Household site soil pollution

Figure: 7.10: Growing vegetables in raised bed gardens

Figure 7.11: Herbicide/pesticide pollution

Figure 8.1: A, B, and C represent the soil profile, a notation firstly coined by Vasily Dokuchaev (1846–1903), the father of pedology. Here, A is the topsoil; B is a regolith; C is a saprolite ; the bottom-most layer represents the bedrock.

Figure 8.2: An illustration of soil pollution

Figure 8.3: An illustration of the world's regions with the most polluted soil

Figure 8.4: An illustration of pesticide application, which leads to soil pollution

Figure 8.5: A summary of the causes and effects of agricultural soil pollution

Figure 8.6: Domestic waste/mixed waste site

Figure 8.7: An illustration of the list of industrial soil pollution causes

Figure 8.8: An illustration of some of the effects of industrial soil pollution, i.e., loss of plant life

Figure 8.9: Dumps of radioactive rock at the Kurung site.

Figure 8.10: How radionuclides are produced during the fission process

Figure 8.11: An illustration of the radioactive soil contamination

Figure 8.12: An image showing the effects of soil pollution by radioactive elements

Figure 8.13: An illustration of chemical extraction

Figure 8.14: An illustration of the agroecological impacts of soil pollution

Figure 8.15: An image showing a brief summary of the possible solutions to the most polluted soil in the world

Figure 9.1: Soil amendments for immobilization

Figure 9.2: Soil Bioremediation

Figure 9.3: The process of waste bioremediation

Figure 9.4: Alternative Strategies for Microbial Remediation of Pollutants

Figure 9.5: Schematic representation of the different steps involved in biodegradation

Figure 9.6: Toxaphene

Figure 9.7: Biostimulation techniques

Figure 9.8: Biostimulation

Figure 9.9: Whey in bioremediation

Figure 9.10: Bioaugmentation

Figure 9.11: Integrated approach of bioremediation including bioaugmentation

Figure 9.12: Handling soil contamination

Figure 10.1: Healthy soils create benefits for people, nature and the climate

Figure 10.2: Soil conservation is essential in ensuring food security

Figure 10.3: The European Green Deal will ensure clean, and sustainable energy through proper soil management

Figure 10.4: The USDA pays farmers to improve their soils

Figure 10.5: Conservation programs prevent erosion and conserve water quality

ABSTRACT

By definition, any compound in soil that surpasses naturally existing levels and poses a risk to human health is a soil contaminant. To give a simple example, lead is a toxic element but exists naturally in certain soils. It was infused in gasoline until 1989, though it is still found polluting soils today. Arsenic also occurs naturally in some soils. But if someone sprays their garden with certain pesticides, it can lead to soil contamination.

This volume examines the main risks of soil contamination in modern areas and previous industrial areas. Typical contaminants in modern soils include pesticides, petroleum products, asbestos, chromated copper arsenate, lead, and creosote. In urban regions, soil contamination is mainly caused by human activities. Some examples include manufacturing, land development, local waste disposal, industrial disposal, and overuse of pesticides or fertilizers. If the soil is contaminated with these compounds, it can damage the natural environment. Most of these compounds are as toxic to flora as they are to people. Additionally, because soil is the "kidney of the planet," pollutants can seep through soil into our water supply systems causing major harm.

PREFACE

Soil pollution, soil contamination or land contamination is soil degradation caused by the existence of xenobiotic (man-made) chemicals or other changes in the natural soil ecosystem. It is usually caused by industrial activities, agrochemicals or improper waste disposal. If the soil is contaminated with these substances, depending on the amount, it can damage the home environment. These elements are just as dangerous to crops as they are to human-beings.

Where and the amount of contamination included in soils largely determines how contaminants spread within an area. The form of soil also plays a key functionin the distribution. For instance, some contaminants may infiltrate groundwater sources more readily than clay. This is due to the quicker infiltration rate of the rough sand type. Fine-grained clay or organic matter in the topsoil can trap pollutants, which means that when left undisturbed(i.e., not excavated or tilled), the pollutants will accumulate.

The volume mentions that some human activities can happen simultaneously. For instance, some new residential complexes are constructed on old industrial venues. These soils can contain buried soil contaminants which are brought up to the surface whenever houses are built or the roads that support them. Emerging pollutants can then be redistributed throughout the modern environment by wind (as dust) or water erosion. Contaminants may also remain on-site, creating on-site exposure risks when food is grown in contaminated soil or when youngsters play in the soil.

Soil contamination triggers a chain reaction. It changes soil biodiversity, reducing soil organic content and the ability of the soil to function as a filter. It likewise contaminates water preserved in soils and groundwater, causing soil nutrient imbalances. The most common soil pollutants include heavy metals, emerging pollutants and persistent organic pollutants - such as personal care items and pharmaceuticals.

Soil contamination is destructive to the environment and affects all life forms that encounter it. Unsustainable agricultural activities that minimize soil organic content promote the transference of pollutants into the food chain. For instance: contaminated soil releases contaminants into groundwater, where these contaminants can build up in plant tissue and then pass on to birds, grazing animals and ultimately humans who consume plants and animals. Pollutants in soil, groundwater, including the food chain cause a wide range of disorders and high mortality in people, ranging from brief acute impacts (such as intoxication or diarrhea) to long-term terminal impacts (such as cancer).

In addition to the environmental impact, soil pollution has high economic costs because of the reduced crop yield and quality. Preventing soil pollution should be the world's top priority. The volume shows that because the vast majority of pollutants come from human activity, this means we have a direct responsibility to make the changes necessary to ensure less contamination and a safer future. The productive capacity of soils and their contribution to food security and maintaining vital ecosystems needs to be recognized and valued.

CHAPTER 1

SOIL AND ITS ROLE IN THE ECOSYSTEM

CONTENTS

1.1. Introduction to Soil as an Ecosystem Service 2

1.2 Medium for Plant Growth 4

1.3 Regulating Carbon and Nitrogen Cycles 6

1.4 Soils and Climate Change 8

1.5 Regulating Water Supplies 9

1.6 Recyclers of Raw Materials 12

1.7 Habitat for Soil Organisms 13

1.8 Engineering and Landscape Medium 15

1.9 Cultural and Physical Heritage 16

1.1. INTRODUCTION TO SOIL AS AN ECOSYSTEM SERVICE

As a natural resource, soil is a vital element in the environment. In the terrestrial ecosystem, soil is highly heterogeneous and has evolved throughout the years. It is the topmost weathering layer of the earth's crust. Its evolution has been brought about by various natural processes. Soil is a habitat for various living organisms and its biodiversity is very broad. There are different types of soil caused by the variability in seasonal temperature, vegetation, parent materials and rainfall. The different types of soils have different mineralogical compositions, chemical and physical properties all of which provide ecosystem services to sustain life. The heterogeneity of soil is said to bring about wide variations of soil reaction to activities causing pollution and causes different impacts that may be specific to a given area(Meli, 2013).

Figure 1.1: Soil is a vital resource in the global ecosystem.

Source: https://www.thespruce.com/easy-diy-soil-tests-2539856

Basic knowledge of soil functions and various aspects of soil resources are useful in understanding interaction of pollutants with soil constituents and their impact on all ecosystems. An ecosystem is defined as a community of organisms and the local environments that they interact with. By studying microbiological processes, soil scientists are able to use ecosystem boundaries to enclose a single soil profile or a single soil horizon. There are various ecosystems on the earth's surface. When nutrient cycling or the effects of management practices on soils are being considered, the ecosystem may be as large as an entire plant community and a soil polypedon system. Pedons are building blocks comprising of soil mapping units and taxonomic classes(Shaner, 2017).

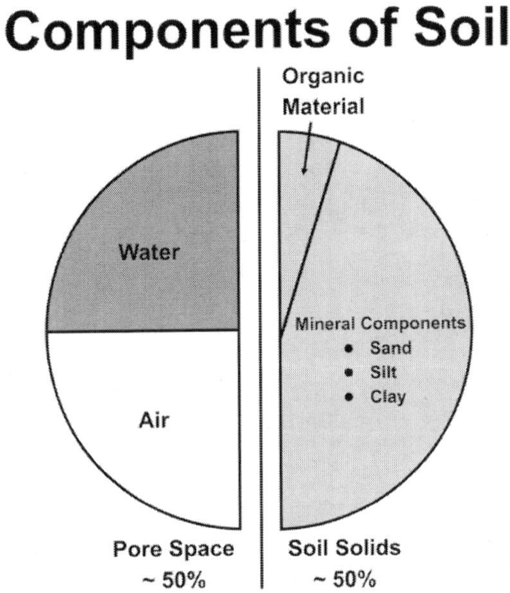

Figure 1.2: Soil contains various components.

Source: https://commons.wikimedia.org/wiki/File:Components_of_Soil_Pie_Chart.png

Soil is very rich as it contains various components including living organisms, air, water, humus and mineral particles. Soil can be grouped into various groups based on various things. For instance, its structure can be used to group soil as healthy or unhealthy. In a healthy soil structure, the particles are arranged in such a manner that the spaces are large enough to allow entry of water and air. The spaces in soil are known as pores. Soil pores allow air circulation in soil facilitating the survival of soil microbes and storing water that will be used by the microbes and the plants themselves (Yakupoglu, Rodrigo-Comino &Cerdà, 2019). Soil pores account for about 50% of soil's volume. In the global ecosystem, soil is useful in regulating natural cycles of mineral matter, organic matter, air and water. Soil is actively involved in the filtering and purification of water. It is also involved in the storing and breakdown of substances. For this reason, it is considered an ecosystem service, wherein there is a constant flow of energy and matter.

A large percentage of the soil's usefulness in ecosystems is attributed to billions of microbes living in soil. In most cases, these microbes and some plants are often viewed with contempt rather than appreciation (Rao, 2014). Some of the organisms living in soil include woodlice, springtails, worms, fungi, algae and bacteria. The most commonly known inhabitant of soil is the earthworm. Formation of soil structure is largely attributed to soil biota. Substances found on the soil surface and in soil are considered useful resources for fodder and food production, basis for forest protection, and source for raw materials. Soil performs various ideal functions as a component of landscapes and nature, as research specimens, religious sites and as geographical and cultural sites. As we analyze soil and its role in the ecosystem, we get to learn that soil is able to fulfill most of its economic and ecological functions when its moisture regime and aeration are not disrupted and there is enough room for plant roots, soil biota and nutrients. Another condition is that there should be a balance of the various living organisms in soil and pollution levels are maintained within various amounts that are tolerable for soil and plant organisms.

As human activities affect soil thereby affecting the global ecosystem, it can easily be pointed out that these activities make soil processes such as soil formation very scarce as most of the processes take a significant amount of time. Soil is considered an inert medium that takes longer durations for them to respond to exogenous influences. For this reason, various problems related to soil are identified later which are difficult to fix. Soil is very delicate and the manner in which it is handled will be reflected by the global ecosystem (Webb, Herrick, &Duniway, 2014).

There are various functions of soil in global ecosystems. These functions are usually a reflection of the functions of soil in other ecosystems across the world. There are five main roles of soil in the global ecosystems including soil serving as a medium for plant growth, regulator of water supplies, recycler of raw materials, habitat for soil organisms and landscaping and engineering medium.

1.2 MEDIUM FOR PLANT GROWTH

As a medium for plant growth, soil plays various roles. Soil provides anchorage for plants. It firmly holds the plant such that it is not destroyed by mild external forces. As an anchorage, it forms the foundation for the plant. The roots of the plant penetrate into deep levels of soil such that the plant is properly supported. Soil also acts as a water holding tank for plants. As plants grow, moisture is very essential as it is involved in the formation of food needed to ensure plant growth. Soil holds water in between its particles. As the roots of plants penetrate in soil, they take up the moisture and transport it to the needed parts. Soil also gives plants a hospitable environment allowing them to take root (Van Leeuwen et al., 2019). There are various properties of soil that affect their growth. They include the soil's ability to hold water, its permeability or aeration, porosity, aggregate soil and soil texture. The kind of soil aggregate size determines the soil type thereby influencing the type of plants that can be planted in that given area. Soil also has humus that is broken down by soil microbes providing nutrients needed by plants. The physical properties of soil will generally influence plant growth. Insufficient moisture in topsoil will cause plant roots to penetrate deeper into soils or the plant will exhibit growth problems.

Figure 1.3: Soil anchors crops to the ground.

Source: https://southeastagnet.com/2020/10/08/assistance-helps-farmers-anchers-conserve-natural-resources/

Plant growth is greatly influenced by the soil's ability to store and supply nutrients to plants grown in it. This is generally termed as soil fertility. Soil fertility is greatly and directly influenced by the organic matter and clay content in soil. Soil fertility is high when there are greater organic matter and clay. In most cases, soils with great organic matter and clay content tend to be dark in color. Plants thrive in such soils (Tsoraeva et al., 2021). By facilitating plant growth, soil is therefore involved in food and biomass production. Soil gives all the needed elements needed to facilitate the healthy growth of plants.

1.3 REGULATING CARBON AND NITROGEN CYCLES

Soil can be said to be central to the earth's ecosystem as it is a dynamic and open habitat with the ability to provide plants with air, nutrients, water and physical support needed for growth. There are various species of animals living in soil as soil is suited to sustain the large populations of animals. All of these organisms are useful in the recycling of chemical elements in soil. Among these animals are fungi and bacteria. They are useful in recycling nitrogen and carbon element that appear toxic to human beings. The nitrogen and carbon cycle are natural processes (Tully et al., 2015). Both cycles involve the uptake of nutrients from soil, restoration of organic matter into soil by the aging and death of tissues, involvement of soil microbes in the decomposition of organic matter and the release of nutrients into soil and the cycle continues. The decomposition of organic matter may involve toxins or nutrients being cycled within the microbial community. Both the carbon and nitrogen cycle are closely linked to the hydrologic cycle. This is because water usually takes up the role of a primary medium in chemical transport.

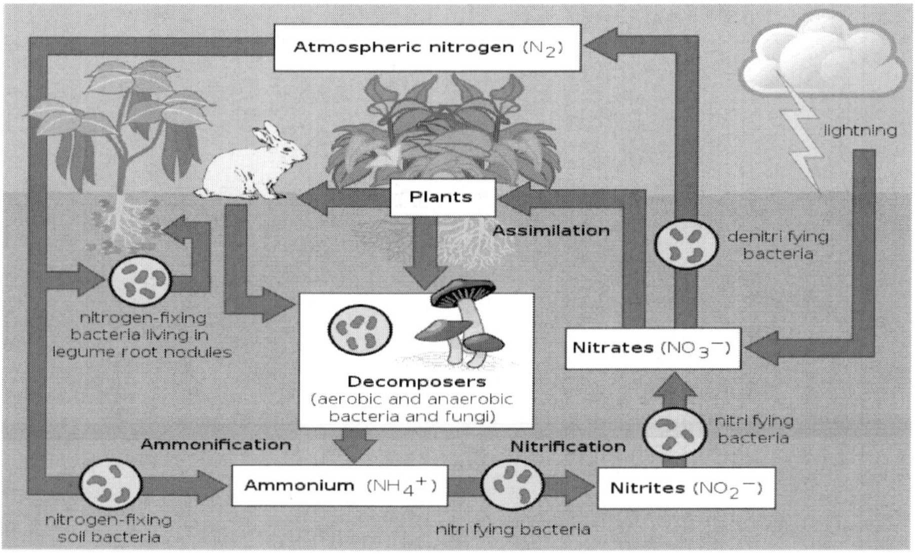

Figure 1.4: Soil helps maintain a balance of carbon and nitrogen in the atmosphere.

*Source: By File:Nitrogen Cycle.svg: *Cicle_del_nitrogen_de.svg: *Cicle_del_nitrogen_ca.svg: Johann Dréo (User:Nojhan), traduction de Joanjoc d'après Image:Cycle azote fr.svg.derivative work: Burkhard (talk)Nitrogen_Cycle.jpg: Environmental Protection Agencyderivative work: Raeky (talk)* *derivative work: Hattiel - This file was derived from: Nitrogen Cycle.svg:, CC BY-SA 3.0, https://commons.wikimedia.org/w/index.php?curid=80960223*

The major component of the atmosphere is nitrogen (78% of the air in the atmosphere). Nitrogen originates in the atmosphere. It finds its way into the ecosystem by biological processes and water cycles. Under biological processes, nitrogen is fixed in soil through nitrogen-fixing organisms in soil. Nitrogen fixation mostly involves nitrogen entering the biosphere as a wet deposition to the soil surface. In such a state, there are various organisms competing for it including nitrifiers, microbial decomposers and plants. Nitrifiers are soil microbes that convert ammonium to nitrate. The amount of nitrogen reentering the atmosphere through the cycle is greatly determined by the extent to which various organisms compete for it (Telles, Guimarães, &Dechen, 2011).

Like nitrogen, carbon originates from the atmosphere and enters the ecosystem from the atmosphere. There are various forms of carbon with the most common being carbon dioxide. During photosynthesis, plants take up carbon and convert it into biomass. Studies show that soil organic matter in the form of biomasses and humus contains about three times as much carbon as land vegetation. Carbon is stored in various forms depending on the region. For instance, in arid and semiarid areas, carbon is stored in inorganic chemical forms such as calcium carbonate. The global carbon cycle is greatly dependent on pools of carbon in various ecosystems. This is attributed to decomposition, its subjection to erosion and its location on land surfaces. It is estimated that soils release about 4 to 5% of their carbon into the atmosphere in a year (Ettler, 2016). This is achieved by the transformation of organic matter into carbon dioxide gas. The transformation process is known as respiration. The amount of carbon dioxide produced from soil respiration is 10 times larger than the amount of carbon dioxide produced when fossil fuels such as petroleum and coal are burnt. After its release into the atmosphere, it is returned to soil as organic matter through the production of biomass.

However, human activities are greatly affecting soil carbon pools by reducing their availability. The natural balance between the release of carbon by soil respiration and the production of carbon-containing biomass are affected by changes in land-use linked to agriculture. An estimate suggested that the imbalance causes an annual net release of carbon dioxide to the atmosphere from agricultural soils to equate to about 20% of the annual release of carbon dioxide from burning fossil fuels. Soil organic matter in temperate regions can reduce as a result of agricultural practices. The reduction could be about 20 to 40% of the original content (Salesa&Cerdà, 2020). The loss may be attributed to soil erosion. The increased flux of carbon to the atmosphere as carbon dioxide also contributes to the loss.

1.4 SOILS AND CLIMATE CHANGE

Climate and soil bear a relation either directly or indirectly. Soil and demographics are affected by the increase in temperature due to global warming and changes in rainfall patterns. The greenhouse effect drives climate change as it causes an increase in levels of gases such as nitrous oxide, methane and carbon dioxide in the atmosphere. Large tracts of land in humid tropics have been converted into agriculture. This has greatly affected the emission of greenhouse gas. It is anticipated that the emission of methane and other gases will affect future climate change. Soil processes contribute about 70% of methane and 90% of nitrous oxide in the atmosphere and can also function as repositories for the gases (Rodrigo-Comino, 2018). There is a source-repository relationship in soil. The processing of methane by soil is greatly affected by the application of nitrogen-containing fertilizers.

Figure 1.5: Placing land under cultivation contributes to the greenhouse effect.

Source: https://www.123rf.com/photo_26430212_cultivated-land.html

Soil plays a vital role in regulating the amount of carbon dioxide in the environment. The term net primary production is used to refer to carbon stored in terrestrial plants through photosynthesis. The net primary production is very important as it is considered the most dominant source of feed, fiber and fuel. About 61 billion tons of carbon are stored in terrestrial plants with forests occupying a significant amount. As several trees have been cut over the years for various purposes, an imbalance has been created. Reforestation of the entire planet to primordial levels may temporarily counterbalance the effect on the release of carbon into the atmosphere. Carbon litter is made from carbon from terrestrial biomass that has not been directly used. Carbon litter is usually incorporated into soil as humus. Other activities such as abandonment of agricultural land, decomposition of wood products, reforestation and deforestation affect the natural cycling of carbon.

1.5 REGULATING WATER SUPPLIES

Soil plays a key role in the regulation of water supplies across the world. It can be considered as nature's water manager. With the changes in climate, there is a great need to properly manage resources such as water. In weather conditions where snow or rain fall on land, they are absorbed by soil. Soil then stores the water as moisture in between the particles between it. Some amounts of water are taken up by plants as they germinate and also small living organisms in soil. However, a large amount is stored in deeper layers of soil (Rodrigo-Comino et al., 2018). By storing water, soil creates a pool of water available for plants, living organisms and human beings to use. Much of this water is utilized between precipitation or irrigation events. When heavy rains are experienced, soil becomes very wet and eventually attains saturation. Once soil attains saturation, water moved downwards into deeper layers through the soil profile. The rate of the downward movement is affected by the size of soil particles. Significant amounts of water move downwards though some are drawn from deeper layers of soil towards the surface through plant transpiration and evaporation.

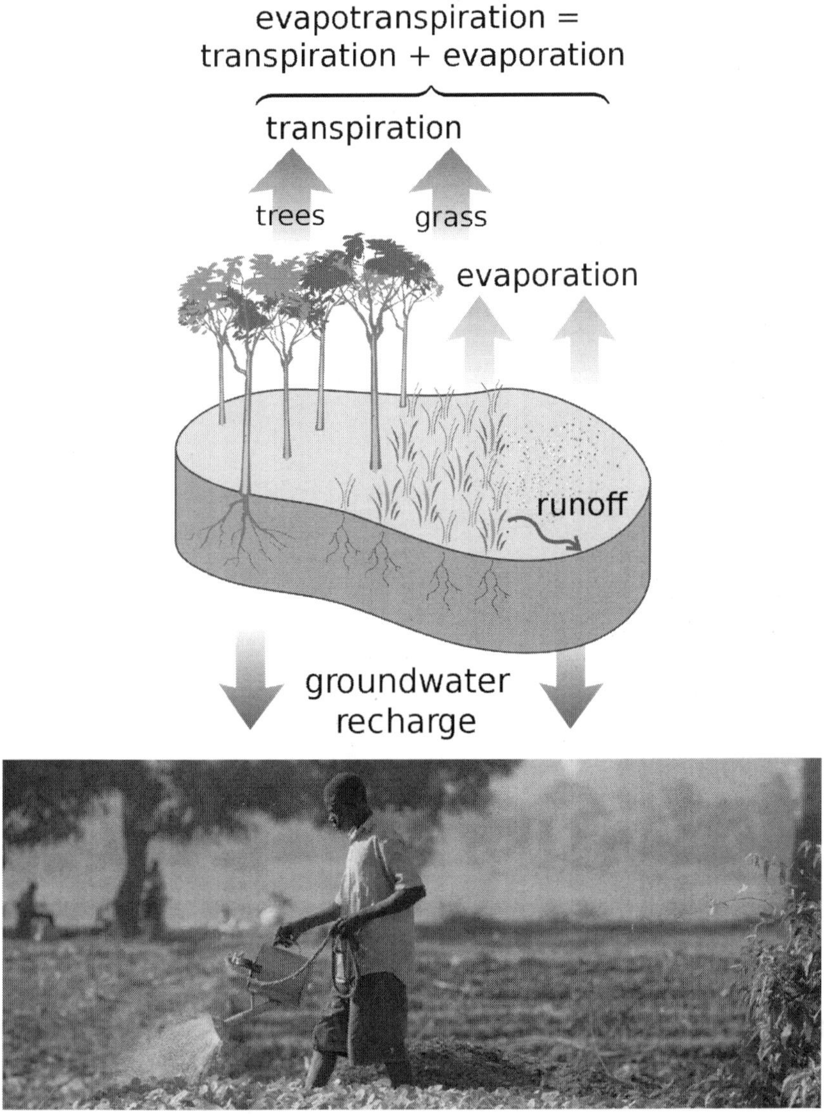

Figure 1.6: Soil enables the preservation of water for future use.

Source: By M. W. Toews - Own work, CC BY 4.0, https://commons.wikimedia.org/w/index.php?curid=2843655

Regulation of water supplies by soil usually comes at a price as most times, soil is at a risk of erosion when exposed to impacts of runoff and raindrops. This really affects water management. The ability of soil to retain and absorb water is reduced as a result of surface erosion. The term water holding capacity is used to refer to the amount of water soil can retain against the pull of gravity. Water holding capacity is largely influenced by the number of very small micro-pores in soil. It is also affected by capillarity which is the movement of water into upper surfaces. Infiltration refers to the rate of water movement into the soil. Infiltration is greatly influenced by soil texture, the soil's physical conditions such as tilth ,soil structure and the amount of vegetation found on the soil surface (Rickson, 2014). For instance, sandy soils have large pore spaces in between them allowing rapid movement of water to deeper layers. However, they are unable to store large amounts of water. Fine textured soils such as clay soil have several micropores. This attribute enables them to retain large amounts of water. The large amounts of water retained cause such soils to have slow rates of infiltration. Water retention abilities are improved by the presence of organic matter. Organic matter also increases infiltration rates of fine textures soils.

When soil has been compacted it becomes rock. Generally, boreholes for water are drilled below the vadose zone (which will include the soil layer, susceptible to pollution) into the underlying aquifer. The soil/vadose layer in general plays an important role in adsorbing and releasing water to the water table as well as regulating pollutants so that they are adsorbed rather than reaching the water table (Kowalska et al., 2018). This enables continuity of life despite harsh climates. Regulation of water supply falls under environmental interaction. Other components falling under environmental interaction include water loss, utilization, contamination and purification. Soil is involved in filtering, buffering and transforming materials between the atmosphere, water table and plant cover. Soil actively interacts with various ecosystems facilitating the decomposition and transformation of waste materials into new materials. Soil naturally has an action that allows filtering. The filtering process allows the capturing of contaminants through soil particles (Poesen, 2018).

Soil particles determine the kinds of contaminants that will be captured. The filtering process causes water to come out cleaner in rivers and aquifers. Soil has the ability to accumulate large amounts of carbon in the form of soil organic matter. This leads to the reduction of the total concentration of carbon dioxide in the environment thereby mitigating global climate change.

1.6 RECYCLERS OF RAW MATERIALS

There are various ways in which soil is involved in the recycling of raw materials. They serve as recyclers in the carbon and nitrogen cycles. In several ecosystems, soil recycles materials through decomposition. Recycling of raw materials is among the greatest functions of soil in global ecosystems. Recycling involves the decomposition of organisms, plants and animals by soil fauna and flora such as insects, fungi and bacteria. The decomposition of these organisms results in the transformation of their remains into simpler mineral forms. The simpler mineral forms are later on consumed by other microorganisms, animals and plants that then make new living tissues and soil humus. The rate of decomposition is influenced by a number of factors such as soil physical environment and the chemical make-up of decomposing materials (Pijl et al., 2020). The physical environment includes the presence of oxygen and moisture in soil that facilitates the survival of soil microbes involved in the decomposition process. Soil temperature also influences the activity levels of decomposing organisms. The amount of nitrogen in soil is categorized under the chemical makeup of materials. The presence of nitrogen in soil affects the digestibility of materials by soil organisms. Faster rates of decomposition are exhibited when the material has high amounts of nitrogen.

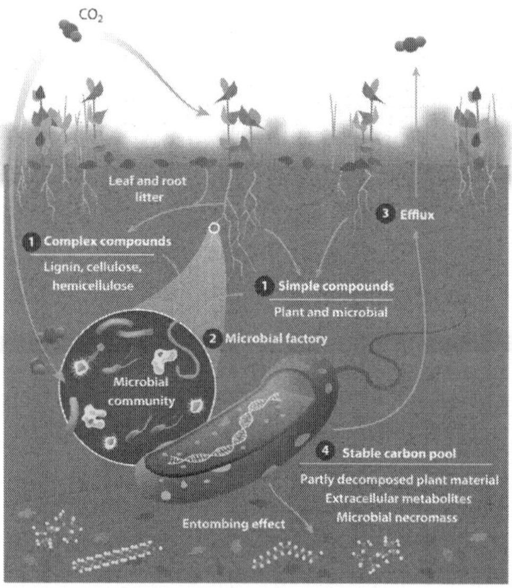

Figure 1.7: Decomposed materials are used by plants to bring forth new life.

Source: https://upload.wikimedia.org/wikipedia/commons/1/19/Soil_carbon_cycle_through_the_microbial_loop.jpg

Soils have the needed capacities to store large amounts of essential plant nutrients and atmospheric carbon. These capacities are created through the humus formation process and also decomposition. The introduction of carbon in soil by microorganisms is considered a biological process. Biologically active carbon is able to remain in soil organic matter for very long durations of time. Carbon sequestration is defined as the temporary storage of carbon in organic matter of soils and biomass. The global carbon cycle is maintained by soil organic carbon. Several studies are being conducted to get information on how land management practices influence soil organic matter levels (Pham et al., 2018). This is because several researchers speculated that land management levels have the potential to impact the occurrence of global climate change. Recycling of raw materials ensures that plants are able to access the needed nutrients. Soil microorganisms also benefit from the recycling process. These processes also allow the fixation of other elements such as nitrogen into soil.

1.7 HABITAT FOR SOIL ORGANISMS

Soil is a habitat for various living organisms of different sizes. Sizes of microorganisms range from very large some of which can be seen. Plant roots also fall under this category. Tiny organisms such as small mites and insects live in soil. The range also extends to microscopically small organisms such as fungi and bacteria. The relationship between soil and soil microorganisms may be symbiotic as soil formation is influenced by activities of soil microorganisms while soil microorganisms get food and water from soil. Microorganisms living in soil are primary soil decomposers. Their main role is to transform and recycle old, dead materials into raw materials utilized by plants during their growth. Among the organisms living in soil are earthworms. As they burrow into soil, they excrete wastes on the soil surface. The waste is broken down by bacteria and other organisms. Different organisms living in soil will consume and digest the organic matter in soil.

Figure 1.8: There are various kinds of organisms living in soil.

Source: By Malcolm Fowles - Own work, CC BY-SA 4.0, https://commons.wikimedia.org/w/index.php?curid=75908673

For the organisms to survive in soil, they need certain basic elements including water, food and air. A suitable physical environment is needed by decomposers in soil. Such a habitat is needed to allow them do various activities. In as much as water is needed by microorganisms, it is not entirely depended upon to ensure their survival. This is because these microorganisms are able to live in harsh environments by staying in a dominant state till water supply is restored. A significant number of organisms living in soil need oxygen and are referred to as aerobic organisms. They include microorganisms and plant roots. Some species of animals have undergone some changes that allows them to survive in areas that do not have oxygen (Pietro & Mahajan, 2022). These animals are known as anaerobic organisms. Microorganisms in soils can breathe properly due to the presence of good soil porosity and large diameters of soil pores. Habitats of soil microorganisms are influenced by soil textural types. Large numbers of soil microorganisms are mostly found in finer soils as they tend to have large numbers of soil micropores. Certain organic material is required by soil organisms to enable them to get carbon and energy. This generally shows that soil microorganisms need food to survive. Large populations of soil microorganisms are found when there are abundant supplies of fresh organic materials.

1.8 ENGINEERING AND LANDSCAPE MEDIUM

Various structures seen on the earth's surface such as buildings, homes and roads are built using soil. Soil also forms a foundation where other structures are set upon them. Soil has various physical properties of soil are useful in building activities. Most of these properties are concerned with construction and engineering properties. They include shrink-swell potential, shear strength, consistency, compressibility and bearing strength (Priori et al., 2020). All of these engineering variables are said to influence various basic soil physical properties such as water content, clay mineral type, structure and texture. Depending on the profile, different soils are used in construction. Sandy soil is mostly used as it has good drainage capability. Poor drainage may cause a building to collapse. Sandy soil is known to easily mix with other construction tools. Soils also contain various elements that can be used in buildings such as marbles. Clay soil is mostly removed before building foundations are established.

Figure 1.9: Soil is used in building infrastructure.

Source: https://theconstructor.org/geotechnical/backfilling-types-procedure-foundation/1648/

Soil has several landscape applications ranging from roadway and bridge construction around highways to courtyards and greenspaces around commercial sites. Soil is also used in lawns of residential housing development. Soil has landscape applications that include planting to improve the beauty of residential areas. During construction, both ecological and physical functions of soils are taken into consideration. However, the extraction of soil to be used in construction negatively affects the environment. Some of the negative effects of soil include soil erosion caused by wind or water. The eroded soil pollutants finally find their way into reservoirs and ponds (Pimentel & Burgess, 2013).

When handling soil during construction, some of the activities undertaken include soil mapping which is the identifying, describing and delineation on maps of different types of soil on the basis of direct field observation or indirect inferences. Some sources of information include aerial photographs. Soil maps are useful in depicting soil properties and functions in the context of certain soil functions such as civil engineering considerations, environmental protection and agricultural food production. These maps are useful in depicting functional interpretations of certain properties such as levels of heavy metals and critical nutrients. Usually, mapping of function specific soil properties is an extension of soil survey. Soil maps on soil components are used alongside auxiliary information in depicting inferences on the specific performance of soil mapping units.

1.9 CULTURAL AND PHYSICAL HERITAGE

Soil has both physical and cultural heritage in various communities across the world. In various communities, soil has more cultural functions. This is because they form part of the cultural landscape of the physical world and also our minds. Cultural attribute is the sense of place and attachment to home soils that have been developed very strongly among certain groups of people. Soil is known to be in existence since the earth was created (Nearing, Lane, & Lopes, 2017). For this reason, it could be a determinant in the migration of people across the world. Other than that, soil acts as a natural earth cover, which protects and preserves physical artifacts of the past allowing us to understand our cultural heritage. Soil has also been used in identifying areas where people settled. Soil is also an essential resource for human productivity. Some rocks found in soil are packed with minerals that can be consumed by human beings and animals. Minerals from these rocks are useful in helping mothers of humans and animals to bring forth new life.

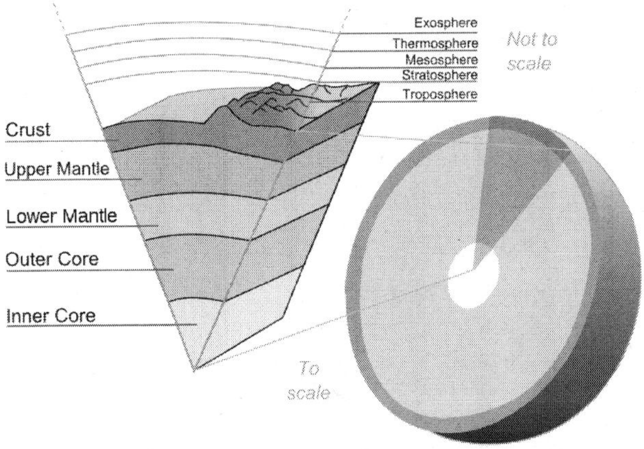

Figure 1.10: Soil is the uppermost cover for the continental land mass.

Source: By Surachit - This W3C-unspecified vector image was created with Inkscape. CC BY-SA 3.0, https://commons.wikimedia.org/w/index.php?curid=2584710

In conclusion, soil is very important in various ecosystems as it ensures the survival and the continuity of organisms living in it and on the earth's surface.

CHAPTER 2

SOIL POLLUTION

CONTENTS

2.1 Soil Contamination Vs. Soil Pollution 20

2.2 Man-Made Pollutants 22

2.3 Common Soil Contaminants of Concern and their Fate in The Soil Environment 26

2.4 Processes that Cause Soil Contamination 30

2.1 SOIL CONTAMINATION VS. SOIL POLLUTION

The term "soil contamination" has frequently been used as a synonym for soil pollution. The Intergovernmental Technical Panel on Soils (ITPS) under the Global Soil Partnership (GSP) has formalized definitions of the two terms (FAO and ITPS, 2015). Soil contamination occurs when the concentration of a chemical or substance is higher than would occur naturally but is not necessarily causing harm (Van Straalen, 2002). Soil pollution, on the other hand, refers to the presence of a chemical or substance out of place and/or present at a higher than the normal concentration that has adverse effects on any non-targeted organism. The soil is one component of the Planet that we seldom see due to the rise of concrete structures and roadways. It goes by various names, including dirt, mud, and ground. It is, nonetheless, quite vital to us. The plants that nourish and sustain us grow on soil, and preserving its health is critical to preserving the planet's beauty. Pollution affects soil, just as it does all other aspects of nature. Soil contamination is a regular phenomenon these times, and it occurs as a result of the existence of man-made materials. Soil pollution is described as the occurrence of hazardous compounds (either pollutants or contaminants) in soil at quantities high enough to endanger human health and/or the environment. Even though the amounts of contaminants found naturally in soil are not significant enough to constitute a danger, soil pollution is defined as when the quantities of contaminants in soil surpass the concentrations that ought to be occurring naturally (Nigussie et al., 2015). It is a massive environmental issue because it contains numerous health dangers. Exposure to soil having high levels of benzene, for instance, raises the risk of acquiring leukemia.

Figure 2.1: In Senegal, a pastoralist herds his cattle past a mound of garbage.

Source: https://reliefweb.int/report/world/soil-pollution-hidden-reality

According to Environmental Pollution Centers, soil pollution is, "The existence of toxic chemicals (pollutants or contaminants) in soil at levels high enough to endanger human health and/or the ecosystem." Even if the amounts of contaminants found naturally in the earth are not significant enough to constitute a danger, soil pollution is defined as "when the quantities of contaminants in soil surpass the concentrations that should inherently be present."

The prevalence of man-made trash is the primary source of contamination in the soil. The waste created by nature, such as dead plants, animal corpses, and rotting fruit and vegetables, only contributes to the soil's fertility. Our waste materials, on the other hand, are filled with chemicals that aren't present in nature and cause soil pollution (Molla&Sisheber, 2017).

Every soil, either polluted or not, consist of a variety of naturally occurring substances (contaminants). Metals, inorganic materials and salts (e.g. phosphates, carbonates, sulfates, nitrates), and many organic molecules are examples of pollutants (such as lipids, proteins, DNA, fatty acids, hydrocarbons, PAHs, alcohols, etc.). These chemicals are mostly generated by soil microbial processes and organism degradation (e.g., plants and animals). Various substances also enter the soil through the environment, such as rain water, as well as from surface water systems and shallow groundwater moving through the soil, and also via wind action or other sorts of soil perturbations. Pollution occurs when the concentrations of soil contaminants surpass natural levels (what is naturally found in distinct soils). Anthropogenic as well as natural causes are the two main sources of soil pollution.

It's critical to realize that all soils include substances that are dangerous or toxic to humans and other living things. The proportion of such compounds in unpolluted soil, on the other hand, is low enough that they really do not represent harm to the ecosystem. The soil is said to be contaminated when the level of one or more harmful compounds is high enough to harm living creatures.

Figure 2.2: Changed Soil properties due to pollution.

Source: By Muntaka Chasant - Own work, CC BY-SA 4.0, https://commons.wikimedia.org/w/index.php?curid=81939788

Soil pollution is caused by a variety of factors that exist every day, if not every minute.

2.2 MAN-MADE POLLUTANTS

Anthropogenic soil pollution is caused by a variety of processes, some of which are intentional (industrial) and others which are not. Soil pollution generated by humans can combine with natural processes to raise hazardous contamination concentrations in the soil.

- Chemical spills and leaks that occur accidentally during storage, transportation, or use (e.g., gasoline and diesel leaks and spills at petrol stations);
- Foundry activities and industrial operations involving furnaces or other procedures that may cause pollutants to disperse in the environment;
- Mining activities that involve crushing and processing raw materials, such as heavy metals, which produce harmful chemicals (Beilicci, Beilicci & Stefănescu, 2014).
- Activities related to construction

- Agricultural activities that include the spread of herbicides, pesticides, insecticides, and fertilizers;
- Transportation operations, which result in the release of harmful vehicle emissions
- Unintentional or planned dumping of chemical waste, like illegal dumping;
- Landfill waste disposal, as waste items may leach into groundwater or produce toxic fumes.
- Broken paint chips, particularly lead-based paint, cascading from building walls.

Let us take just a few examples from these man made soil pollutants and extrapolate

Figure 2.3: Man made soil pollution.

Source: By Fquasie - Own work, CC BY-SA 4.0, https://commons.wikimedia.org/w/index.php?curid=120649455

2.2.1 Construction

Owing to their near-ubiquitous character, construction sites are by far the most significant sources of soil contamination in urban settings. Almost any chemical used on building sites has the potential to damage the soil. Chemicals that can easily flow through the air as fine particulates, on the other hand, pose a greater risk. PAHs, for example, are more difficult to degrade and bioaccumulation in living beings because they travel as particulate matter (Blake et al., 2021).

Furthermore, building dust can rapidly travel through the air and is particularly harmful due to its smaller particle size (less than 10 microns).

Construction powder has been linked to respiratory disorders like asthma and bronchitis, as well as cancer. Furthermore, demolition projects that feature older structures may release asbestos, a harmful material that can contaminate soil. The wind can move around these particles.

2.2.2 Agricultural Activities

Since the invention of modern pesticides and fertilizers, the use of chemicals has increased dramatically. While these soil additives may improve agricultural yield, they contain chemical contaminants that are not easily broken down. As a result, when they mix with water, they pollute the soil and gradually deplete the soil's fertility.

2.2.3 Disposal of Waste

The volume, toxicity and improper management of wastes is an increasing source of concern. Although industrial waste will undoubtedly pollute the environment, there is another way wherein we contribute to pollution. Industrial waste disposal and disposal of diapers are the primary activities resulting in the contamination of soil through improper waste disposal. Each person excretes a certain amount of waste in the form of urine and feces (Rao, 2014).

While the majority of sewage is disposed of in sewers, a significant quantity is deposited directly into dumps in the guise of regular waste (Meli, 2013).Even the sewage system comes to a conclusion at the landfill, where biological waste contaminates the soil and water. This is attributable to the fact that our bodies are filled with toxins and chemicals, which are now pouring into the ground and polluting the soil.

2.2.4 Oil Spills

Oil spills are common during storage and transportation. This is a common problem at gas stations. The toxins in the gasoline degrade the soil quality and render it unfit for cultivation. These contaminants can infiltrate groundwater through the land and render it unfit for human consumption.

2.2.5 Acid Rain

Acid rain is brought about by chemical reactions that start when elements like nitrogen oxides andsulfur dioxide are delivered up high. These substances can ascend exceptionally high into the air, where they blend and mix with water, oxygen, and different synthetic compounds to form more acidic poisons, known as acid rain. Dead or dried trees are a typical sight in regions affected by acid rain. Acid rain drains aluminum from the dirt (Hu et al., 2017). That aluminum might be destructive to plants as well as creatures. Acid rain additionally eliminates essential nutrients and minerals from the earth that trees need to develop.

2.2.6 Industries

Industrial pollution is viewed as a significant element that contaminates the environment. It brings about ecological disruption and forces high costs on society and on human health and safety. Textile, concrete, sugar, tanning, petroleum, glass, and plastic are major polluting sectors (Zhao et al., 2015). Manufacturers pollute air by delivering different poisonous gases into the air like carbon monoxide. Smoke is produced by chemical and paper plants, treatment facilities and so forth. The processing of petroleum derivatives additionally contaminates air.

2.2.7 Deforestation

Though mining causes deforestation at a lot more limited scope than horticulture, it creates high quantities of air and water contamination that taints environmental conditions. When trees are chopped down, quite a bit of that captured carbon is delivered into the climate again as CO2. This is the manner by which deforestation and forest destruction add to global warming (Ettler, 2016).

2.2.8 Urbanization

The environmental contamination brought by urbanization is mostly due to the impacts, for example, the increment of metropolitan land use, of the extreme convergence of populations. The damage of the environment through economic development is chiefly because of the excessive utilization of energy. Urbanization additionally brings about deforestation, habitat loss, and the extraction of freshwater from the ecosystem, which can diminish biodiversity and change species ranges and interactions.

2.2.9 Mining

Many lots of rock are uncovered, moved, and squashed in mining activities fundamentally increasing how much residue and particulates is released in the air. What's more, mine tailings, which might contain finely ground and harmful material, can become airborne (Shaner, 2017). This air contamination can directly influence human health. Mining negatively influences the ecosystem by instigating loss of biodiversity, soil erosion, and pollution of surface water, soil and groundwater.

2.3 COMMON SOIL CONTAMINANTS OF CONCERN AND THEIR FATE IN THE SOIL ENVIRONMENT

Xenobiotics - compounds that are not naturally occurring in nature but are created by humans – are among the most dangerous soil contaminants. The name 'xenobiotic' comes from the Greek words 'xenos' (foreigner) and 'bios' (life) (life). Several xenobiotics have been identified as carcinogenic. Below is an illustration of the most common soil pollutants.

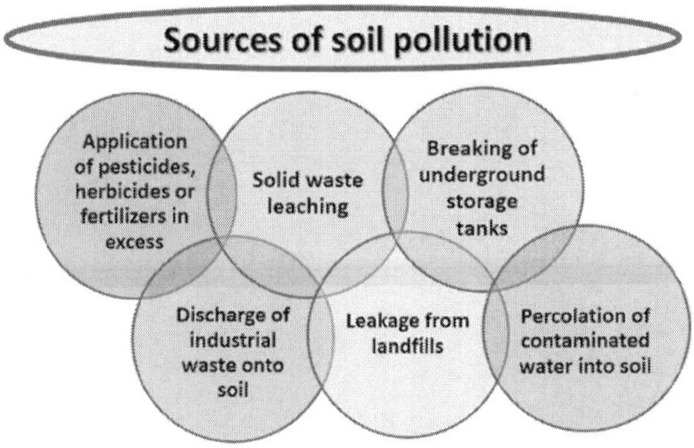

Figure 2.4: Pollutants that Contaminate Soil.

Source: Gavrilescu M. Water, Soil, and Plants Interactions in a Threatened Environment. *Water*. 2021; 13(19):2746. https://doi.org/10.3390/w13192746

2.3.1 Metal/Metalloid

Heavy metals (like lead and mercury, in excessively high amounts) in soil can make it highly poisonous to humans. Below is a list of metals that can be classed as soil contaminants.

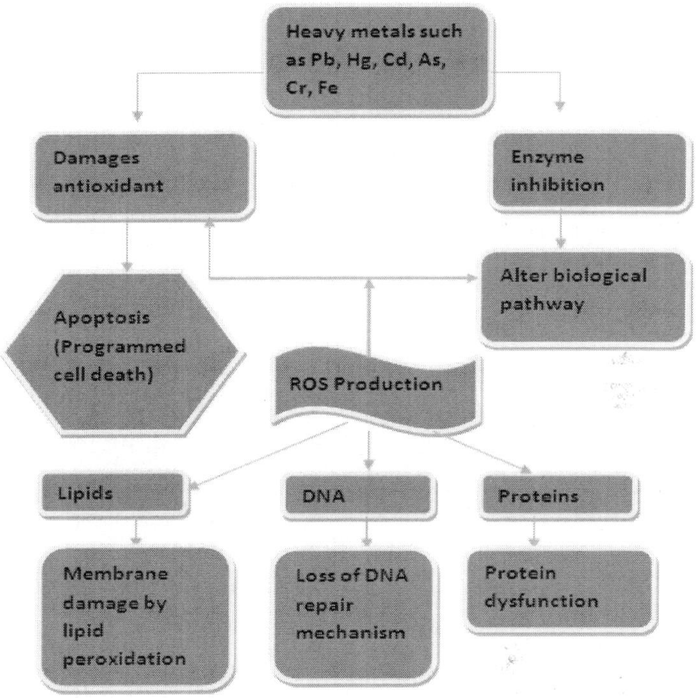

Figure 2.5: Toxic Metals that Cause Soil Pollution.

Source: Anyanwu BO, Ezejiofor AN, Igweze ZN, Orisakwe OE. Heavy Metal Mixture Exposure and Effects in Developing Nations: An Update. Toxics. 2018; 6(4):65. https://doi.org/10.3390/toxics6040065

These metals can come from a variety of places, including mining, agriculture, electronic trash (e-waste), and medical waste.

2.3.2 Nonmetal

Polycyclic Aromatic Hydrocarbons

PAHs (polycyclic aromatic hydrocarbons) are a class of organic chemicals that are characterized by:

1. Only carbon and hydrogen atoms are present.
2. Their chemical structures contain more than one aromatic ring.

Naphthalene, anthracene, and phenalene are all examples of PAHs. Polycyclic aromatic hydrocarbons (PAHs) have been related to a variety of cancers. In humans, these chemical substances can induce cardiovascular disorders (Mekuria et al., 2009).

Coke (coal) production, automobile emissions, tobacco smoke, and shale oil extraction are all sources of PAHs in the soil.

Figure 2.6: Cigarette Butt.

Source: By Sillyputtyenemies - Own work, CC BY-SA 3.0, https://commons.wikimedia.org/w/index.php?curid=6498918

2.3.3 OrganicAromatic

Industrial Waste

Soil pollution can occur when industrial waste is discharged into the environment. The following are some prevalent soil contaminants that can be traced back to industrial waste.

- Industrial solvents containing chlorine
- Dioxins emitted during pesticide manufacturing and trash incineration.
- Plasticizers/dispersants
- Polychlorinated biphenyls (PCBs)

Various petroleum hydrocarbon waste products are produced by the petroleum industry. Some of these pollutants are recognized to be carcinogens, like benzene (Meshesha et al., 2012).

2.3.4 Pesticides

Pesticides are chemicals (or mixes of chemicals) that are used to kill or prevent pests from reproducing. Pesticides employed in agriculture include:

- Herbicides - used to eliminate weeds and other undesirable plants in agriculture.
- Insecticides - used to kill insects or inhibits them from engaging in undesirable or destructive behaviors. • Fungicides — chemicals that kill or slow the growth of parasitic fungi.

However, inadvertent pesticide spread (also known as "pesticide drift") causes a slew of environmental issues, including water and soil pollution. The following is a list of some of the most common pesticide pollutants foundin soil.

2.3.5 Herbicides

- Triazines
- Carbamates
- Amides
- Phenoxyalkyl acids
- Aliphatic acids

Figure 2.7: Herbicides application.

Source: https://cdn.britannica.com/26/143726-050-8570DDED/herbicides-cereal-crops.jpg

2.3.6 Insecticides

- Organophosphates
- Chlorinated hydrocarbons
- Arsenic-containing compounds
- Pyrethrum

2.3.7 Fungicides

- Mercury-containing compounds
- Thiocarbamates
- Copper sulfate

Humans are exposed to a variety of health concerns as a result of these pollutants. Pesticides can cause disorders of the CNS, immune system problems, cancer, and congenital abnormalities, to name a few. (Ma, Zhao, & Zhu, 2021).

- Aside from the rare occasions where natural chemical accumulation results in soil pollution, natural cycles could also have an impact on human-released hazardous chemicals in the soil, reducing or increasing pollutant potency and/or soil contamination levels. This is owing to the intricate soil environment, which includes the existence of other chemicals and natural circumstances that may interplay with the pollutants produced.

Natural processes leading to soil contamination:

- Natural chemical accumulation in soil as a result of imbalances between atmospheric deposition and evaporation with rainfall water
- Soil natural contamination under specific environmental circumstances
- Contaminated rain water leaks into the subsurface (Rao,2014).

2.4 PROCESSES THAT CAUSE SOIL CONTAMINATION

Soil contamination can be generally split into two categories –

- Naturally produced soil contamination
- Anthropogenic soil contamination (caused by human activity)

2.4.1 Natural Contamination of Soil

Some contaminants naturally build in soils as a result of relatively unusual processes. This can happen as a result of the atmosphere's unequal deposition of soil. Another way for this type of soil contamination to arise is by the transport of soil contaminants with precipitation water (FAO, 2022).

The buildup of substances having the perchlorate anion (ClO4–) in some drier, arid habitats is an illustration of natural soil pollution (Rao et al., 2007).It is crucial to remember that some pollutants might be created naturally in the soil as a result of particular environmental conditions. During a thunderstorm, for example, perchlorates can occur in soils holding chlorine as well as certain metals (Dongdong et al., 2022).

2.4.2 Anthropogenic Soil Pollution

Almost all incidents of soil pollution are caused by humans. Soil contamination can occur as a result of a range of human activities. Some of these processes are given below.

- Asbestos contamination of surrounding soil may occur during the demolition of ancient buildings.
- The use of lead-based paint throughout building projects can also poison the soil with dangerous lead concentrations.
- Spillage of fuel and diesel in transit can pollute soils with petroleum hydrocarbons (Blake et al., 2018).
- Metal casting factory (foundry) activities frequently promote the spread of metallic pollutants into surrounding soils.
- Underground mining operations can lead to heavy metal contamination of land.
- Inappropriate disposal of extremely toxic industrial and chemical waste can lead to serious soil pollution. The accumulation of toxic wastes in landfills, for example, can result in waste seepage into the soil. This garbage has the potential to harm groundwater (Kowalska et al., 2018).
- Pesticides contain a number of potentially toxic compounds. Pesticides used in an inefficient or excessive manner might lead to serious soil pollution.
- Sewage from densely populated regions can contaminate soil. Several carcinogenic chemicals may be present in these wastes.

- Nuclear waste, e-waste, and coal ash are examples of waste that can harm soil.

Pollutants and contaminants are found in soil contamination. Biological hazards and some anthropogenic impacts are the primary soil contaminants (Meli, 2013). Soil contaminants are all the byproducts of soil pollutants that pollute the soil. Human activities that poison the soil range from pesticide-infested agriculture practices to urban or industrial pollution or radioactive discharges that pollute with a variety of harmful substances.

Figure 2.8: Different Sources of pollution.

Source: Sui X, Wang X, Li Y, Ji H. Remediation of Petroleum-Contaminated Soils with Microbial and Microbial Combined Methods: Advances, Mechanisms, and Challenges. Sustainability. 2021; 13(16):9267. https://doi.org/10.3390/su13169267

CHAPTER 3

IMPACTS OF SOIL CONTAMINATION

CONTENTS

Introduction.......34

3.1 Human Health Impacts.......34

3.2. Ecotoxicological Impacts on Terrestrial Ecosystems.......36

3.3. Impacts on Aquatic Ecosystems.......37

3.4. Food Chain Disruptions.......37

3.5. Air Quality Impacts.......37

3.6. Groundwater Contamination.......38

3.7. Socio-Economic Impacts.......38

3.8. Soil Pollution: Possible Solutions.......39

3.9 Causes of Soil Pollution.......40

3.10 Industrial Accidents.......57

3.11 Intensive Farming.......59

3.12 Petrochemicals.......62

3.13 Road Debris.......65

3.14 Drainage of Contaminated Surface Water Into the Soil.......68

3.15 Waste Disposal.......71

INTRODUCTION:

Soil has an impact on practically every area of our daily life. We don't always understand it. As a consequence, we often fail to see the impact of soil pollution on our daily life. Polluted soil leads to stunted crops and, in some cases, a toxic subsurface water table. Soil contamination has a wide range of negative repercussions on flora, fauna, humans, and the entire ecosystem. Polluted soil presents a higher threat to children since they are more prone to infections. This section discusses some of the most significant effects of soil pollution (Li et al., 2018).

3.1 HUMAN HEALTH IMPACTS

Given that soil is the reason we can feed ourselves, its contamination has serious ramifications for our health. Crops and plants grown in dirty soil absorb a large portion of the pollution and pass it on to humans. This could account for the abrupt increase in minor and fatal ailments.

Long-term contact with such soil can change the genetic design of the body, resulting in congenital disorders and chronic health issues that are difficult to treat. In fact, it can sicken livestock to a significant level and contaminate food over time. Soil pollution can potentially cause widespread famines if plants cannot thrive in it (Chuchkalov, Fadeev, & Alekseev, 2020). Contaminants in the soil can be present in all three periods. As a result, these contaminants can enter the human body through a variety of routes, including direct skin contact or inhalation of polluted soil dust.

Short-term (acute) impacts on human health from contaminated soil include:

- Headaches, nausea, and vomiting
- Coughing, chest discomfort, and wheezing
- Itching of the skin and eyes.
- Tiredness and weakness

Soil pollution has been connected to several long-term (chronic) illnesses. The following are some examples of such disorders (Hu et al., 2017).

- Prolonged exposure to high concentrations of lead can cause lasting nervous system harm. Children are more susceptible to lead poisoning.
- Central Nervous System (CNS) Depression.

- Injuries to internal organs like the kidney and liver.
- An increased risk of developing cancer.

3.1.1. Environmental Impacts

Soil Degradation and Loss of ecosystem services

a. Impacts on Biodiversity

Because soil contamination is frequently accompanied by a reduction in nutrient availability, plant life is unable to grow in such soils. Plants can be poisoned by polluted soil with inorganic aluminum. Furthermore, this form of pollution frequently raises the salinity of the soil, making it unsuitable for plant growth (Lal, 2019).

Through a process called as bioaccumulation, plants grown in contaminated soils can collect significant amounts of soil pollutants. All of the contaminants generated in these plants are transmitted up the food chain when herbivores consume them. Many valuable animal species may become extinct as a result of this. Furthermore, these contaminants have the potential to reach the top of the food chain and cause diseases in humans.

Many soil contaminants, including petroleum hydrocarbons and industrial solvents, have been related to congenital problems in humans. As a result, soil pollution can have a number of harmful implications on human health (Cerda et al., 2018).

The extensive pollution of the soil has an impact on the ecological equilibrium of any system. When the composition of the soil changes dramatically in a brief span of time, most vegetation will be unable to adjust. The fungi and bacteria in the soil that hold it together start to deteriorate, causing soil erosion as a result.

The soil's fertility gradually declines, rendering the land unfit for cultivation and preventing any local vegetation from surviving. Large swaths of land have become contaminated, posing a health risk. Unlike deserts, where native vegetation thrives, such land is unsuitable for most types of life.

b. Loss of soil organic carbon and soil nutrients (soil fertility)

Soil degradation describes what happens when soil quality and its ability to support plants and animals decline. Soils may lose certain physical, chemical or biological properties that underpin their internal network of life (Zhao et al., 2015). While soil degradation is a natural process, it can also be caused by human activities. Intensive agricultural practices such as deforestation, overgrazing, intensive farming, wildfires and construction have accelerated land degradation in recent decades.

c. Degradation of soil structure and increase in soil erodibility

Soil degradation describes what happens when soil quality declines, reducing its ability to support plants and animals. Soils may lose certain physical, chemical or biological properties that underpin their internal network of life. Degraded soil cannot support life forms and is easily eroded away in case of natural disasters like floods.

3.2. ECOTOXICOLOGICAL IMPACTS ON TERRESTRIAL ECOSYSTEMS

Terrestrial ecotoxicology examines how environmental pollutants impact terrestrial organisms and their environment. It requires three components: (a) source, (b) receiver, and (c) route of exposure. Contaminants enter the terrestrial ecosystem through direct application, diffuse sources, or long-term transportation. Terrestrial receptors comprise of soil microorganisms, invertebrates, mammals, amphibians, reptiles, birds,and plants. Terrestrial organisms may be exposed to contaminants through skin, oral cavity, inhalation and food chain exposure. Two key processes determine the movement of pollutants in and through terrestrial food webs: bio-magnification and bioaccumulation (Van Straalen, 2002).

Understanding the relative significance of these processes and the different routes of exposure can help to understand potential toxic effects. Toxicity responses from lad organisms can be evaluated using standard laboratory toxicity testing protocols; there are no standard methodologies for reptile toxicity testing.

Nevertheless, most species are not directly tested, but use a standard number of test organisms from which other species are inferred. For some chemicals, the amount of contaminants present in plant or animal tissue can be applied to predict whether they will be affected (Shaner, 2017). Bioassays can be performed in the field to study possible effects under natural factors. Population models may be applied to determine whether toxic impacts resulting from exposure to environmental contamination significantly alters the capacity of populations to sustain themselves over time. Pollutants can also alter the composition of plant or animal species in a community.

3.3. IMPACTS ON AQUATIC ECOSYSTEMS

Soil pollution affects above-ground and below-ground biodiversity in two key ways. Organisms can be reduced in numbers due to contaminant-induced toxicity, however, community-level changes can also be observed that benefit sensitive or resistant organisms over organisms that are sensitive to contaminants. It has also been observed that exposure to certain trace elements leads to resistance in soil microorganisms and also promotes antimicrobial resistance (Heydari, 2020). Additionally, soil contaminants can infiltrate the food chain and cause disease and death of soil, terrestrial (including humans) and aquatic life forms. Thus, loss of biodiversity and biomass can lead to a reduction in organic matter and variations in nutrient input and cycling. This compromises the primary productivity of natural and agricultural ecosystems and leads to a general loss of soil ecosystem services.

3.4. FOOD CHAIN DISRUPTIONS

According to a report, soil pollution adversely affects food security in two ways: it reduces crop yields due to toxic concentrations of the pollutant, and crops grown on contaminated soil are unsafe for animal and human consumption. Loss of organisms at higher nutrient levels, for instance, carnivores, can also break food chains. In kelp forests, sea urchins comprise the main consumers of kelp. Sea otters feed on sea urchins. However, sea urchins can destroy kelp forests when sea otter numbers are reduced by disease or hunting. Soil pollution has dire consequences: it affects food security, both by reducing crop metabolism, thereby reducing crop yields, and by making crops unsafe for animal and human consumption (Rao, 2014).

3.5. AIR QUALITY IMPACTS

Air pollution can seriously affect the quality of soil and water resources. When polluting our air, we also pollute the precipitation that falls into soil and water bodies.

Air pollution can destroy plants and trees in a number of ways. Ground-level ozone may lead to reduced crop and commercial forest yields, lower sapling growth and vigor, and increased crop susceptibility to disease, pests, and other environmental stressors such as severe weather (Meli, 2013).

Air pollution can directly contaminate water surfaces and soil. This can destroy crops or minimize their yields. It can destroy young trees and various plants. Acid rain occurs when sulfur dioxide and nitrogen oxide particles in the atmosphere mix with oxygen and water in the atmosphere.

3.6. GROUNDWATER CONTAMINATION

Soil pollution also contaminates the subsurface water table. Toxins inside the soil could simply permeate slowly and persistently into the water table because this water is held under the layers of soil.

We must also keep in mind that this is the water that can be obtained from wells and tube wells for consumption and use. When such poisonous water is eaten or utilized over time, it has a number of negative consequences for human health (Kelly et al., 2020). Long-term ingestion of this hazardous subsurface water causes diseases such as arsenic poisoning, food poisoning, and others. These illnesses have the potential to be lethal.

3.7. SOCIO-ECONOMIC IMPACTS

3.7.1. Social Impacts

The social impact of soil contamination should not be underestimated. Research has identified factors that make living in polluted communities especially stressful. These comprise a lack of individual control over the event, as well as confusing, insufficient or conflicting information about contamination and the potential for lasting harm or damage. Pollution has been found to strain marital relationships, lead to increased divorce rates, and strain other relationships within families and with friends, relatives and colleagues.

Affected people often experience depression, helplessness, and constant worry about their own and their children's health. Parents often mention feeling guilty about raising their children in regions they believe could harm their long-term health. Research has also revealed links between stress and unhealthy behaviors, including smoking, alcohol consumption, illicit drug use, poor diet, and antidepressant prescriptions (Hu et al., 2017). Community issues, such as open conflict, are also highly reported, especially when there is strong disagreement over possible health risks or issues of resettlement and compensation.

3.7.2. Economic Impacts/Costs

The quantifiable economic losses from soil pollution include reduced soil productivity and crop yields, food contamination and loss of sales, loss of biodiversity and water quality. Poverty and weak political and legislative systems are often seen as the main hindrance of least developed countries in their response to greener and more sustainable growth (Zhao et al., 2015). However, past experience shows that poorer countries are not completely incapable of addressing soil pollution, but rather often lack the technical capability and technology to do so. This makes knowledge and technology transfer from more developed countries critical to addressing socioeconomic disparities caused by environmental degradation, albeit knowledge transfer needs to be aligned and merged with indigenous and local knowledge for community acceptance.

3.8. SOIL POLLUTION: POSSIBLE SOLUTIONS

Soil pollution is a complicated issue that must be addressed. It is critical that we all recognize the importance of soil to our survival. The sooner we recognize the problem, the easier it will be to rectify the issue of soil pollution. Because it is a complicated situation, everybody, from a person to the government, must work together to solve it.

To address soil remediation, several remedial options are available. The following are some major ways for decontaminating contaminated soil (Kheyrodin, 2016).

- Pollutant removal through heat remediation — the material is heated to drive volatile organic contaminants into the vapor state, where they can be recovered via vapor extraction.
- Bioremediation, also known as phytoremediation, is the use of microbes to decontaminate soil
- Mycoremediation is the use of fungi to accumulate heavy metal pollutants.

3.9 CAUSES OF SOIL POLLUTION

3.9.1 Microplastics

Microplastics are a major cause of soil pollution, and can potentially alter the entire balance of ecosystems. They can be termed as plastic particles smaller than 5 mm. Anexample of this is the microbeads. Microbeads in personal care products, are responsible for 50% of global plastic contamination in marine environments (Kumarasinghe, 2021). Many of the household items we use on a daily basis contain plastics. Microplasticsare tiny enough to pass through our water treatment plants,as these microplastics migrate into our soils they make their way into crops and livestock feed.

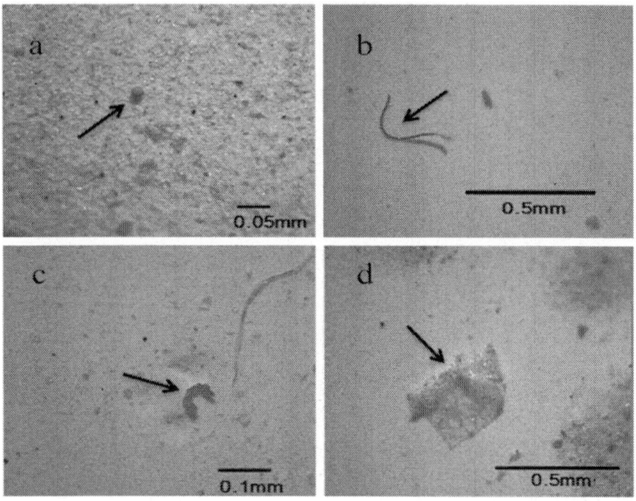

Figure 3.1: Sample elements of micro-plastics.

Source: Li Y, Zhang Y, Chen G, Xu K, Gong H, Huang K, Yan M, Wang J. Microplastics in Surface Waters and Sediments from Guangdong Coastal Areas, South China. Sustainability. 2021; 13(5):2691. https://doi.org/10.3390/su13052691

As plastics break down, microplastics are released into the environment. This is especially true in urban areas where microplastics have been found in larger quantities, particularly in soil contaminated with sewage. Microplastics are really, really bad. So small they can be overlooked by environmental regulators, microplastics contaminate our water, soil and bodies without us even realizing it. Unbreakable, nondegradable, and inedible: we have made a monster.

Other sources of microplastics include vehicle tires, clothing and fishing nets. Harmful effects of microplastics include habitat destruction and ingestion. Microplastics enter the environment through multiple pathways such as sewage effluent, urban storm-water runoffs and washing of synthetic clothes (Kavian et al., 2019). They are extremely detrimental to the environment and life on earth. However, the relationship between microplastics and the ecology of natural systems is still poorly understood.

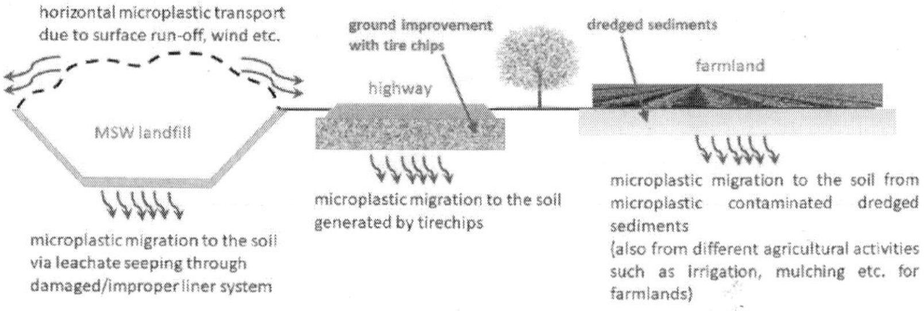

Figure 3.2: The environmental fate and impact of microplastics on soil.

Source: Monkul MM, Özhan HO. Microplastic Contamination in Soils: A Review from Geotechnical Engineering View. Polymers. 2021; 13(23):4129. https://doi.org/10.3390/polym13234129

Microplastics are sometimes dumped directly on the land surface or disposed of in landfills that sometimes spill over onto the land surface. In many instances, this improperly disposed waste causes damage to the surrounding environment whether it is open land or watercourses that eventually flow towards the sea. The effects of microplastics do not physically orient our individual views on how they impact soil. Instead, their harmful effects on animals and humans contribute to one of the leading causes of soil pollution (Jomaa et al., 2013). Whereas other forms of plastic pollution tend to wash back out to sea or break down into smaller pieces over many years, tiny pieces of plastic travel deep into soil, where worms and bacteria consume them alongside organic matter in the dirt. Then these microplastics travel up the food chain as worms are consumed by birds and other larger animals.

Ever wondered what happens to the plastic bags and bottles you throw away? Well, they don't magically disappear. Plastic doesn't degrade easily so it can take up to a million years for a plastic bottle to decompose. Also, plastic bags take 10-20 years just to decompose in landfills and even longer in the ocean. Microplastics contribute to one of the leading human causes of soil pollution. Scientists estimate that over 300 billion tons of microplastics currently contaminate the world's soil. Microplastics have been found in nearly all types of soil, both urban and rural. Microplastics not only contaminate the soil, but are harmful to soil microbes and plants. It is estimated that if nothing is done to mitigate these inputs by 2035 there will be more microplastics in the soil than organic matter. The good news is that there are many responsible companies who are working with governments to address this growing crisis.

Microplastics from clothes, cosmetics and other household products are everywhere. It's hard to know how much damage these are doing to our soil as microplastics make their way into the food chain (Issaka& Ashraf, 2017). It is also important to note that the effects of microplastics today are rather detrimental and similarly do have a negative impact on future generations. The problem of microplastics is rather complicated. But the solution starts with us humans. Reducing microplastics is one of the most important ways we can help the environment.

Figure 3.3: Using magnets is one way of removing microplastics while mitigating their effect on soil.

Source: Magalhães S, Alves L, Medronho B, Romano A, Rasteiro MdG. Microplastics in Ecosystems: From Current Trends to Bio-Based Removal Strategies. Molecules. 2020; 25(17):3954. https://doi.org/10.3390/molecules25173954

A new focus on the effects of microplastics on soil is providing a critical missing link that addresses one of the most neglected aspects of microplastic pollution. Plastics are part of our everyday lives, and scientists have found them in some surprising places (Ighodaro, Lategan, & Yusuf, 2013). But we know that many of the items littering our roads, beaches, landscapes and oceans will make their way into the ground where their impact extends far beyond aesthetics. Something that everyone can do to help reduce microplastics in the environment, including soil, is to simply stop using plastic items that you throw away. Use a reusable straw, bag, or water bottle instead.

3.9.2 Oil Spills

While there are many other causes of soil pollution, about 10% of them come from oil spills. Oil spills are a cause of soil pollution because they introduce contaminants into the natural environment (Van Straalen, 2002). Oil spills happen either on land or in water, and the effects vary depending upon where they occur. Oil often kills plants and tiny organisms such as bacteria, worms and insects that live in soil. The chemicals in oil can break down toxic substances and bacteria in water, which results in reduced oxygen levels in water bodies that often suffocate aquatic life.

Figure 3.4: Oil spills are among modern man's problem in the care of soils.

Source: https://mharefinery.com/possible-restoring-of-land-damage-by-oil-spill/

While it caters to 10% of the total threat towards soil, oil spills are the biggest threat to soil pollution and can cause irreparable damage to a large area in one spill event. In addition to soil contamination and loss of property, many species of animals, birds, fish and plants die from oil spills as well. Consequences of soil pollution include a decrease in healthy food production and an increase in the levels of toxins in the human body, leading to health problems. Typically, only large oil spills in marine environments make news headlines. The cover-up and removal of traces are usually difficult and quite costly. Spilling oil into the soil can cause long-term problems. If a large amount of oil is spilled due to an accident on land or while loading onto or off of a ship, it can contaminate the soil in a large area around the spill site (Haregeweyn et al., 2015). There is a great concern about oil spills as they can produce significant and long-lasting adverse effects on terrestrial and marine environments. The first step in managing threats to human health and welfare is risk assessment, which develops an understanding of the environmental and public health hazards associated with chemicals.

Oil spills can result from human activities. Oil spills can contaminate the soil and can lead to disastrous explosions and fires. Government and industry players are developing standards, procedures and regulations to reduce the risk of accidents and spills, and to clean up when spills occur. Oil has the capacity to damage terrestrial habitats, especially those that are fragile or protected, such as mangroves, and destroy the natural ecosystems that constitute these environments (Kowalska et al., 2018).

Figure 3.5: Oil spill on the land has an immediate impact on soil.

Source: By US Gov NOAA - US Gov NOAA, Public Domain, https://commons.wikimedia.org/w/index.php?curid=10905

Oil leaks and spills result in contaminated soil and ground water, at times reaching and polluting the underground aquifers. The most common sources of oil pollution in the environment are older, underground oil storage facilities and automobile use. Cutting down these would significantly decrease oil spills, which are a leading cause of pollution of the soil. Avoiding soil pollution caused by oil spills is an area where the public sector has been able to make considerable progress in the past few years (Ha, Van Dung & Ngoc, 2013).

The soil that is affected by oil spills is irreversibly damaged, they do not get absorbed or washed away. The oil gets mixed with the soil and affects the soil's texture and fertility adversely. Oil spills have long-term ill effects on the soil which results in damaging its composition and leaving it unfit to be used for growing crops or plants. Oil spills at sea, from ocean-going vessels, offshore platforms, drilling rigs and wells can cause major environmental harm through the release of hydrocarbons into the sea. The oil can be found floating on the surface of the ocean together with other spill-related waste such as containers and structural materials resulting from a collision or storm. Oil can also be found below the surface in layers or droplets depending on the type and specific gravity of oil and the distance it has dispersed from the source. Subsurface oil can eventually reach shorelines where currents carry water in toward land whilst wind-driven surface currents push floating matter away from shore through varying degrees.

3.9.3 Mining and Activities by Other Heavy Industry

Activities from mining and other heavy industries contribute to soil contamination. Poorly managed mining operations can lead to high levels of metals and other harmful substances in the soil (Guerra et al., 2017).

Figure 3.6: Effluents from acid mine drainage: example of Rio Tinto, Spain.

Source: By Carol Stoker, NASA - ACD03-0051-13 from http://www.nasa.gov/centers/ames/news/releases/2003/03images/tinto/tinto.html, Public Domain, https://commons.wikimedia.org/w/index.php?curid=4359761

Mining involves the extraction of heavy minerals, such as iron ore and copper, from geological deposits. This usually occurs at much deeper levels in the subsurface than many other types of land disturbance, so the area directly affected by each mine is generally small. The main environmental impact associated with mining is the alteration and destruction of land around the mine site, which can cause fundamental changes to soil chemistry, physical structure and ecological functioning (García-Ruiz et al., 2017). Mining land is often left with a pit after materials of interest have been extracted. In some cases these pits are allowed to fill naturally with water, creating lakes. Mining operations can also change drainage patterns and fill soils with sediment. The remaining waste is stored in above-ground landfills known as tailings impoundments. These facilities must be carefully constructed to prevent pollutants from escaping into the environment when water has infiltrated them.

These deposits form a mineralized package that is of economic interest to the miner. Ores recovered by mining include metals, coal and oil shale, gemstones, limestone, chalk, dimension stone, rock salt, potash, gravel and clay. Mining is required to obtain any material that cannot be grown through agricultural processes, or feasibly created artificially in a laboratory or factory. Mining in a wider sense includes the extraction of any non-renewable resource such as petroleum, natural gas and even water. Modern mining processes involve prospecting for ore body's analysis of the profit potential of a proposed mine extraction of the desired materials and finally reclamation of the land to prepare it for other uses once the mine is closed.

In the course of its development, agriculture has always been the primary cause of soil erosion and degradation. However, since the industrial period, it has been catching up with other types of pollution (such as air and water) that have their source in industies (Frankl et al., 2016). The build-up of polluted dust on soils is attributed to wind erosion. These properties make them highly suitable for industrial purposes; however, there are also accompanying disadvantages to using metal oxides as catalysts.

The mining industry has been a controversial one since its inception. Not only does mining provide the materials we need to build roads and buildings, extract valuable minerals, etc., but it can cause a significant negative impact on the environment and the health of workers. The major problem associated with this form of degradation is that once polluted; remediation becomes very expensive since pollutants are often found at depths below which vegetation will grow naturally or most crops do not tolerate poor soil conditions.

Heavy metals contribute to soil pollution. Polycyclic Aromatic Hydrocarbons (PAHs) are toxic because when in excess, these chemicals produce harmful effects resulting in carcinogenic, mutagenic and teratogenic effects. These include TNAH, tetraethyl lead, and other hydrocarbons that can severely affect the environment. One of the main causes is pesticides as well as fertilizers used on farms. Another major cause is acid rain which can destroy vegetation, crops, and wildlife. Finally, landfills contribute to soil pollution by mixing in with water systems thus killing off aquatic life.

Heavy industrialization is known to have negative impacts on soil composition. It can cause the soil to become imbalanced due to the introduction of large amounts of chemicals and waste matter. As a result, the soil is made infertile and may lead to the death of organisms (Fenta et al., 2021). Soil pollution also causes land deterioration in which the soil becomes dry and hard.

3.9.4 Accidental Spills during Activities

Spills and leakage from leaking pipelines, from improper handling or disposal of toxic or hazardous wastes, or from fertilizer and pesticide application can be major direct sources of soil contamination. These are not accidental. They are intentional activities with very specific desired outcomes, such as increased crop yield or killing/inhibiting pests. Accidental spills during activities such as excavating for foundations, digging for utility lines and sewer lines, improperly covered waste piles during construction, or drainage of polluted water over the soil surface can also cause soil contamination (Eskandari et al., 2022). The oil from gun cleaning or paint thinner from painting can easily seep down into the soil if spilled on the ground.

Figure 3.7: The adverse effects of accidental spills.

Source: By Kallol Mustafa - Send via Email to Rahat Rahim, CC BY-SA 4.0, https://commons.wikimedia.org/w/index.php?curid=37383332

Accidental spills can occur during any type of activity, including mining, shipping, manufacturing, storing, and transportation. There were two cases of accidental contamination of the soil with fuel oil in 2007. The first occurred when a leak developed due to corrosion of the outer surface of a fuel oil pipeline and some 5500 tons of crude oil was spilled over an area of up to 42 ha. The second incident was due to the breakage of electric power transmission towers, which resulted in more than 50 tons of electric insulating oil being spilt over 30 km of roads (Cogo, 2022).

Soil pollution means the presence of substances in the soil that harm the ecosystem and, consequently, human health. Accidental spills during activities can be the cause of soil pollution. These spills may result from the inappropriate handling of hazardous materials or from accidental leaks, for example, because of corrosion. The use of this material poses a risk to groundwater and soil under normal circumstances. For this reason, regulations have been introduced to prevent soil pollution caused by oil exploration, extraction and transportation operations. Today, Accidental spills are a leading cause of soil pollution. Various substances and materials that cause soil pollution include oils, fuels, toxic chemicals, pesticides, detergents and lead from discarded batteries.

Most accidental spills are oil-based, originating from such sources as cars, trucks, and boats. Dumping or leaking may also contribute to soil contamination. Construction site management and green chemistry are two approaches to mitigate Accidental spills during activities. We're all guilty of leaving behind a trail of carelessness and spills during activities. The world has suffered because of our ignorance (Zhang & Huang, 2015). Accidental spills can be prevented by checking that the oil reservoir is not opened unnecessarily during activities and for vehicle checkups.

Human activities that may cause accidental spills include spills from petrol and other automobiles, leaks or spillages from factories in industrial areas, and oil wells. Accidental spills can be dangerous as they poison the soil as well as the crops grown on it. If these spills occur on a farm, this may potentially destroy the soil and all the crops. This will surely be a huge loss for the farmer. One ought to incorporate spill prevention and mitigation strategies into your planning. Spill prevention aids can assist you in preventing spills before they occur. If a spill should happen, containment or mitigation controls can minimize spilled liquid from reaching the land or a waterway. By combining prevention and mitigation controls, the risk of contaminating soil, waterways, or groundwater is reduced.

Processes and strategies are already in place to address the problem, such as reducing the amount of waste chemicals disposed of inappropriately and promoting sustainable farming practices. Other strategies include implementing stricter disposal laws, banning certain pesticides and fertilizers, and recycling when possible. By switching to alternative products that are safer for the environment, people may help reduce the problem (Yakupoglu, Rodrigo-Comino &Cerdà, 2019). Accidental spills during activities are malfunctions or failures of practices or equipment that leads to accidental spills, which can cause soil pollution. Accidental spillages are usually the result of mishandling, misuse, or failure to use equipment, structures and instruments. Accidental spills during activities can cause soil pollution. The above basic prevention measures need to be established in order to mitigate accidental spills during activities.

3.9.5 Corrosion of Underground Storage Tanks

Figure 3.8: A corroded tank in a mine.

Source: https://upload.wikimedia.org/wikipedia/commons/d/d9/Corroded_iron_tank_in_abandoned_mine_on_White_Island.jpg

Corrosion of underground storage tanks (USTs) can lead to the release of tank constituents into soils, which can then cause soil pollution. The above terms can be defined subsequently as; Corrosion (the natural process of converting any substance, including metals, to its simple forms) of underground storage tanks (including piping used to transmit the contents), termed a major cause of soil pollution. Soil pollution from leaking USTs can present significant liabilities for petroleum marketers, property owners and lessees, environmental consultants, and users of USTs (Shaner, 2017). The steel tank is what we call a "conductor" because it conducts electricity with the environment, thereby causing corrosion and making the tank fail.

It is particularly important that those interested in preventing soil pollution through store tank corrosion understand that no coating will withstand the corrosive effects of gasoline, diesel fuel and jet fuel indefinitely (Ettler, 2016). The best practice is to use a composite liner in conjunction with a scrubbing system to remove contaminated water from beneath the liner. Underground storage tanks (UST) are used in shipping and distribution facilities to store fuel or chemical products. A single tank failure may have catastrophic consequences such as the escape of chemicals into the environment or fire.

The impact of USTs also vary depending on the cause. Corrosion of underground storage tanks can occur due to electrochemical action and/or liquefaction caused by moisture or having a pH or electrical potential between 0.3 and 15. Any metal tank will corrode at high enough temperatures. Yet indeed, whether due to chemical attack or electrochemical reactions with the surrounding soil, is a major source of soil pollution, especially in areas where fuels and chemicals are stored. Corrosion of underground storage tanks (USTs) and piping used to transmit the contents has been identified as a major cause of soil contamination in the United States. USTs are often constructed out of steel owing to their strength, availability, and low cost. However, steel and many other metals can corrode under certain environmental conditions and lead to leaks in USTs and associated piping (Van Leeuwen et al., 2019).

Rain, condensation or water runoff can gather in the tank and cause what's called a float valve, which is a valve that automatically adds or removes fluid to maintain the desired fluid level. Heavy rain or flooding can overwhelm such underground storage tanks and cause siphoning of toxic material into water supplies on neighboring properties. It has been reported that 3% of underground storage tanks were compromised by corrosion in the past decade and there are also cases where contaminated soil was disturbed by an excavator and adjacent buildings were damaged (Rao, 2014). In many instances, waste cleaning projects involving the removal of underground storage tanks (USTs) may require coordination with environmental agencies and/or departments of motor vehicles (DMVs), which regulate USTs in some states.

Petroleum products, metals, and other compounds can leak into the soil and become concentrated in certain areas. This is particularly true when there is a large accumulation of disposed materials, such as may be found in a landfill or around an industrial site. Other reasons for soil pollution include urbanization and agricultural practices (Tsoraeva et al., 2021). Urbanization commonly causes the buildup of impervious surfaces that may lead to increased runoff or flooding during storms. Agricultural practices also cause soil pollution if they are not properly managed to prevent erosion and the use of pesticides or fertilizers in excess of what the environment can release safely.

Factors that affect the rate of corrosion include construction materials, tank design and maintenance, installation methods, configuration and contents. The above effects can be mitigated if those tanks are lined, are emptied and cleaned before closing them, use no pipe fittings made of non-ferrous metals, comply with all applicable local, state, and federal requirements regarding the closure of such facilities. However so many have termed it impossible to avoid corrosion in underground steel storage tanks including piping used to transmit their contents. Therefore the protection of tanks against corrosion is imperative. Measures that may help in protecting storage tanks against corrosion include buried storage tank installation techniques, the use of modern coatings and cathodic protection.

Maintaining the physical integrity of these storage tanks is a problem where there are numerous challenges to achieving success. These fixtures are subject to corrosion from a variety of causes, but primarily from exposure to soil acids and microorganisms. There is also potential for localized corrosion and discharge of the stored material into neighboring soil strata (Tully et al., 2015). Each of these causes can be mitigated by proper methods during construction, operation, and maintenance procedures that are properly adhered to both in the short term and long term. The best way to mitigate the risk of underground storage tank corrosion is to use (1) corrosion-resistant materials and construction, (2) cathodic protection, and (3) an outer barrier. In addition to these options, for extra safety, you can also maintain an environmental engagement plan, which includes monitoring and mitigation steps in case of a leak.

Protection from corrosion is essential to avoid future contamination. In addition, tanks storing substances with low pH or highly reactive chemicals are at risk of corrosion. Corrosion protection for USTs is regulated to reduce the amount of damage that can occur when leaks and spills occur.

3.9.6 Acid Rain

Acid rain causes pollution to water, soil and air, which in turn causes damage to plant life, wildlife and human health and infrastructure. It is a form of rain that is overly acidic due to normal rain becoming contaminated by other pollutants in the air due to factory emissions, gas vapors and other pollution sources (Meli, 2013). Acid rain causes higher soil acidity, which affects the ability of trees, forest crops and plants to obtain nutrients such as magnesium, calcium and nitrogen from the soil. Acid rain is rain that contains nitric and sulfuric acid. Snow and fog can also contain nitric and sulfuric acid. When nitrogen oxides and sulfur oxides are released into the air from burning fossil fuels, these pollutants react with water, oxygen, and other chemicals to form sulfuric and nitric acid. When sulfur dioxide and nitrogen oxides are released from power plants and automobiles, prevailing winds blow these compounds across state and national borders, sometimes over hundreds of miles (Telles, Guimarães, &Dechen, 2011). Acid rain is a type of acid deposition, which can appear in many forms. Through dry deposition such as dust, particularly harmful pollutants can also be transported to sensitive environments. The pH of acid rain is below 5.6. Normal rain has a pH close to 5.6, which is slightly acidic because carbon dioxide and water in the air react together to form carbonic acid, a weak acid. Acid rain, or more precisely acid precipitation, is the term that describes rainfall, snowfall or fog containing high levels of hydrogen ions. The presence of these ions allows the solution to be acidic due to the fact that hydrogen atoms lose an electron which turns them into a cation. This causes many problems for ecosystems and their inhabitants as well as results in the deterioration of man-made structures.

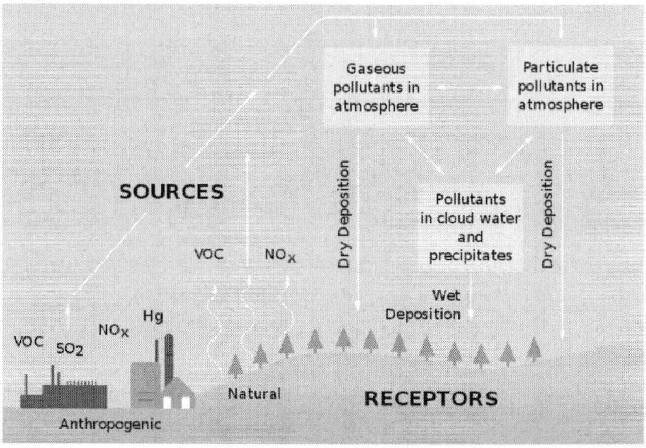

Figure 3.9: The adverse effects of acid rain on soil.

Source: By Origins.gif: The original uploader was NHSavage at English Wikipedia. derivative work: Zazou (talk) - Origins.gif, Public Domain, https://commons.wikimedia.org/w/index.php?curid=7038792

Acid rain can cause structural damage to metal and concrete and erode limestone, marble and soapstone. In soil, acid rain makes aluminum more available for plants to absorb, which may harm trees at high levels. Acid rain, which is acidic when it reaches the earth, gradually affects soil formation. Acid rain penetrates the soil and dissolves certain minerals and plant nutrients such as calcium, magnesium and potassium. In addition, acid rain can liquify aluminum compounds in the soil, which are then easily washed away by subsequent rainfall or runoff (Salesa &Cerdà, 2020). Acid rain, which consists of water containing high levels of hydrogen ions and sulfate ions, can have negative effects on soil. When acid rain falls on soil, the acidity in the soil dramatically increases. This phenomenon can cause metals in the soil to be released into water supplies, which then causes environmental hazards by entering bodies of water and other sources of drinking water. Acid rain is toxic to soil, and it can affect a wide range of soil organisms.

When acid rain leaks into the ground and reaches groundwater supplies, the groundwater carries some of that acidity with it. Plants also absorb this acidic water through their roots, which damages plant cells. The negative effects of acid rain on soil can have far-reaching impacts on the overall health of plants and animals in the ecosystem. Acid rain can also pollute soil at a rate of more than 200 metric tons per year. This acidification of soil means that plants have a much harder time getting nutrients from the soil. Soil is made up of minerals, water, and nutrients. Acid rain also contains water, but the excess acid in it stifles plant growth and releases minerals, like calcium and magnesium, that plants need. Acid rain, when it falls to earth or when fog settles on earth and turns acidic, is highly corrosive (Rodrigo-Comino, 2018). In some soils and bodies of water, acid rain alters the natural pH levels so dramatically that it prevents life from existing in those environments. When acid rain falls on earth, the sulfuric and nitric acids it contains make the soil more acidic. Acid rain does not form directly in the soil, but it can make its way into the soil from other sources: via dry deposition (acid particles falling through the air and settling onto earth), surface runoff (water flowing over soil collecting pollutants as it goes), or groundwater flow carrying pollutants into soil.

3.9.7 Pesticides

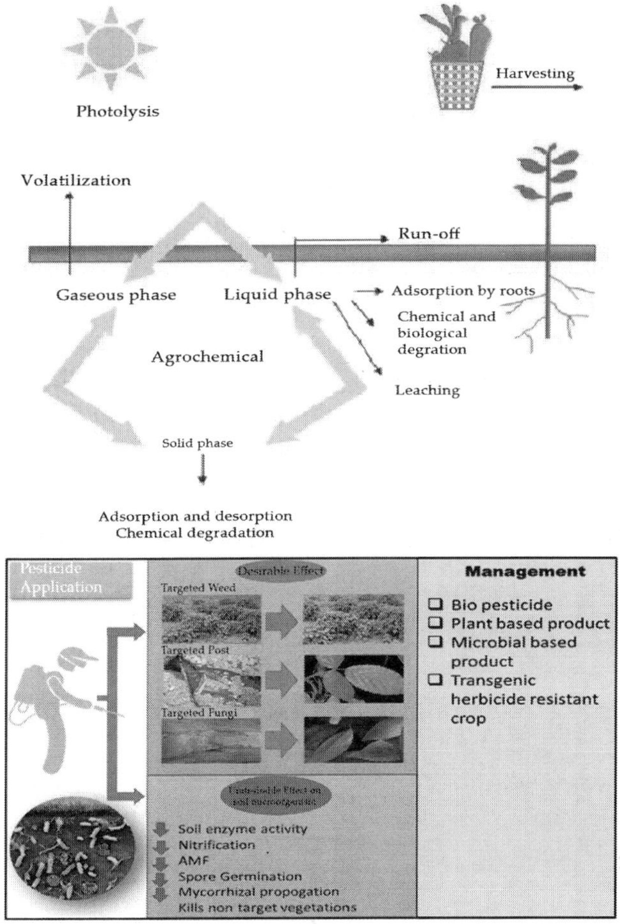

Figure 3.10: An illustration of the effects of agrochemicals on soil.

Source: Meena RS, Kumar S, Datta R, Lal R, Vijayakumar V, Brtnicky M, Sharma MP, Yadav GS, Jhariya MK, Jangir CK, Pathan SI, Dokulilova T, Pecina V, Marfo TD. Impact of Agrochemicals on Soil Microbiota and Management: A Review. Land. 2020; 9(2):34. https://doi.org/10.3390/land9020034

Agrochemicals are a range of widely used chemicals in agriculture for controlling pests, weeds and diseases. When polluting soil, agrochemicals can be taken up by plant roots, translocated within the plant, and accumulated in living tissue in a way that depends on the original soil levels of these pollutants. Harvestable plant parts or their residues can enter the human food chain and, thus, become a potential source of risk for public health. Soil pollution by agrochemicals, although a significant problem all over the world, is much underreported due to economic and political reasons. Agrochemicals are a major source of soil contamination by organic pollutants, especially in rural areas (Hu et al., 2017). Agrochemicals may pollute surface and underground waters, leading to a decline in water quality and biodiversity. Agrochemicals mostly used are pesticides, herbicides and fertilizers.

Pesticides are chemicals that reduce or eliminate pests, including microbes, fungi, weeds, insects and rodents. They were developed to minimize crop loss due to these pests. Pesticides pollute soil and water through runoff and soil erosion because they stay in the environment and take long periods of time to degrade. Soil pollution by pesticides is a form of pollution in which chemicals intended to control pests contaminate soil. When pesticides are used, harmful chemicals can leach into the soil, which can be transported through the environment with water and air (Rodrigo-Comino et al., 2018). Because of this, soils near agricultural areas tend to have higher levels of pesticide contamination than other areas. Pesticide pollution in the soil is largely due to the repeated use of chemical fertilizers. The excessive application of chemical fertilizers can cause the accumulation of pesticide residues in the soil, thereby affecting the degradation rate of pesticides. Pesticide residues are slowly degraded and accumulated. When the residual concentration exceeds a certain level, they may harm people's health or human reproductive function, and even induce cancer. The excessively used chemical fertilizer causes eutrophication in rivers and lakes, and even destroys living creatures in water bodies

Herbicides can pollute soil by a number of different mechanisms. Herbicides are pollutants that can accumulate in the soil. The herbicides used to treat the soil before planting and after harvest can pollute the soil by decreasing its fertility and killing certain organisms. Herbicides represent a major source of soil pollution. The predicted risk of surface water pollution varies from 6–19% for the European Union to 19–46% for China, with the compounds pyridate, diuron and tebuthiuron being the major contributing factors in Europe and isoproturon and mecoprop in China. The growing use of herbicides has resulted in the increasing pollution of soils. Although many studies on the fate and effects of herbicides on soil microorganisms, plants, and animals have accumulated in recent years, little is known about the influence of their presence on biological nitrogen fixation in plant roots.

Fertilizers results in soil pollution. Some fertilizers leach into underground aquifers, polluting our drinking water. Heavy use of chemical fertilizers can have a toxic effect on our soil and surrounding ecosystems. Fertilizer pollution has been shown to damage freshwater and marine ecosystems across the globe. Frequent use of fertilizers can cause high salinity in the soil. This is due to the accumulation of salts and chemicals that are used in producing these fertilizers which causes an imbalance in the nature and acidity of the soil. Fertilizers affect soil negatively but when used in small amounts they can be helpful. Fertilizers are a problem because every person in the United States uses an average of over 500 pounds per year (Rickson, 2014). This level of fertilizer use can cause serious problems since overuse causes pollutants to enter our soil. Fertilizers are essential to keeping lawns looking green and lush, but they do have drawbacks. When used improperly, fertilizers can affect soil negatively by polluting it with high levels of nutrients that promote algae growth in lakes, rivers and streams near you. The addition of fertilizers to soil can drastically alter the environment and make it less fertile. Farmers use fertilizers to increase crop yields, but excess use of these chemicals in the form of nitrogen and phosphorus can lead to a variety of potentially harmful consequences, such as polluted drinking water supplies. Excessive fertilizer use may also contribute to the development of oxygen-poor dead zones in coastal areas and a rise in greenhouse gas emissions.

3.10 INDUSTRIAL ACCIDENTS

The awful soil pollution by industrial accidents can cause severe harm to plants or even kill them. Industrial accidents pollute soil, and as it is very difficult to clean up the polluted soil unlike water and air, it can pollute the plants and trees for a long time. Industrial accidents can take several forms, each producing a threat to soil. Explosion or toxicity of hazardous materials results in the contamination of both soil and groundwater (Poesen, 2018). Soil pollution by industrial accidents can be serious to the environment. Accidental spills, leaks and other unforeseen events can contaminate soil with harmful chemicals that can harm our environment and ecosystem. Accidents are defined as an unanticipated occurrence that causes personal injury or damage to property and must be reported immediately to the state agency.

3.10.1 Ammunition, Chemical Agents and Other Agents of War

Ammunition (informally ammo) is the material fired, scattered, dropped or detonated from any weapon. Ammunition is both expendable weapons (e.g., bombs, missiles, grenades, land mines) and the component parts of other weapons that create the effect on a target (e.g., bullets and warheads). Nearly all mechanical weapons require some form of ammunition to operate. Pollutants released from ammunition into the soil may include lead, copper, nickel, cadmium and zinc. Lead bullets left in the environment can cause soil pollution. Nerve and mustard gas agents, fired in artillery shells or other weapons, left dangerous contamination when they hit the ground. Although the chemical agents themselves degrade with time, the explosives and toxic metals in ammunition, such as lead, barium and mercury do not biodegrade. Depending on the type of munitions and type of soil, munitions residues may be highly mobile or may stick to the soil particles (Pijl et al., 2020).

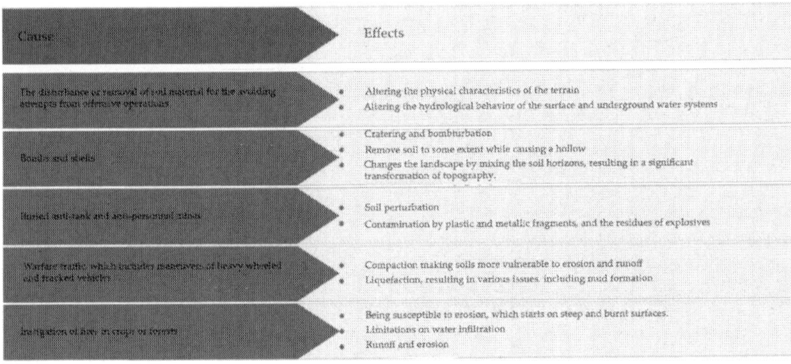

Figure 3.11: Effects of artillery and contamination.

Source: Broomandi P, Guney M, Kim JR, Karaca F. Soil Contamination in Areas Impacted by Military Activities: A Critical Review. Sustainability. 2020; 12(21):9002. https://doi.org/10.3390/su12219002

Soil is a non-renewable resource that can be contaminated by heavy metals from old ammunition. Ammunition pollutes the soil by containing heavy metals. In particular, shot from shotguns and lead bullets are toxic. It is also common for ammunition to contain copper, barium and antimony. This can contaminate a large area of land and affect many animals. The negative effects may be long-term as lead remains in soil for hundreds of years. By contaminating the soil, ammunition has a negative effect on both humans and the environment. Since humans use soil as a medium to grow their food, there is an increased likelihood of food contamination. While the effect of contaminated soil on plants is not yet clear, weathering of ammunition can produce toxic pollutants that can poison nearby soils and water resources. Ammunition can become lodged in the soil causing contamination. Depending on what type of ground, a round of bullets does not disintegrate. Surrounding materials would be affected (Pham et al., 2018). For example, lead left in soil would cause the next batch of vegetables to become contaminated. Given the increasing use of high-caliber weapons and military-grade ammunition, the issue of soil contamination by heavy metals needs to be addressed. Metals in shooting range bullets can contaminate the soil for years. Runoff from bullets can affect local streams and rivers and even seep into groundwater aquifers. Heavy metals in spent ammunition can contaminate soil and groundwater long after the munitions are fired, but new research has shown that plants can help limit the pollution. Phytoremediation may be performed by planting trees or crops. The less expensive "phytostimulation" technology relies on fertilization, irrigation, or both to enhance plant uptake of contaminants from soil.

3.11 INTENSIVE FARMING

Intensive farming is one of the most common cause of soil pollution. In modern agriculture, intensive farming is an approach to crop production characterized by high inputs of labor, fertilizers, and capital that yields high outputs in land products over a short period of time. Although the theory behind it is sound, this approach can cause major soil pollution. The term intensive farming describes farms that use: lots of chemicals, the latest technology, the best seeds and practices to reduce soil erosion (Pietro & Mahajan, 2022).

The agricultural industry has long been known to cause damage to the environment in various forms, chiefly soil pollution. Soil pollution is primarily caused by harmful chemicals, also known as contaminants, being added to the soil, often through agricultural practices. Soil pollution is often caused by farming practices such as over-use of chemical fertilizers, pesticides, use of sewage sludge, industrial waste, etc. These fertilizers tend to degrade the soil structure and make it inhospitable for any life other than weeds and pests.

Figure 3.12: A piece of land bearing the adverse effects of intensive farming.

Source: Image by Everything connects

In the threshold of extensive farming, plants also are a probable cause of soil pollution. Some woody plants, and many ferns and grasses, produce thin but extensive roots in order to extract nutrients and water from a wide area of soil. Different types of plants have grown in different regions because they are adapted to the climate and soil conditions. Plants will grow according to their genetic blueprint until they are affected by their environment. Thus, soil pollution is a problem that affects whole communities because it limits the growing of plants or causes them to grow abnormally.

Intensive farming is far more common in modern society. It has been made possible through modern technology. Modern machinery and techniques such as the use of pesticides, herbicides, insecticides, fungicides, and fertilizers are used to increase the yield. In the case of fruits and vegetables, the rigorous application of technologies such as tissue culture, seed science and nurseries (Priori et al., 2020). Most importantly, mechanization has had a big contribution in this regard. One of the reasons why intensive farming has become so popular is that it is a very productive way to farm. For example, a small area of land can support more cattle using intensive farming practices than if that land were set aside as grazing land.

The aim of organic farming is to mitigate intensive farming as a cause of soil pollution. There is indeed a need for it to be diverted. Understanding the causes and effects of intensive farming on land environments can help reduce soil pollution. To mitigate the problem of soil pollution caused by intensive farming, farmers must be given incentives to rotate crops and not use excessive nitrogen inputs.

The greatest challenge to soil pollution is intensive farming due to the high demands for food, therefore finding alternatives for it. The use of urban farming and vertical farming is beneficial for the fight against soil pollution since it does not demand huge areas and allows the repurposing of land that cannot be used for farming. With the world's increasing population, the demand for locally produced food has resulted in land being utilized for farming many times over. This intensive farming puts more pressure on existing soil and has known adverse effects on soil quality. Farmers must mitigate this pollution by switching to practices that reduce its harmful effects.

When soil is exposed to different forms of pollution, it becomes unusable and uninhabitable for farmers. Modern farming techniques such as intensive farming caused a rapid depletion of soil nutrients, which may cause the soil to become unfit for further farming purposes. A more sustainable practice would be to incorporate a rotating pasture system in which livestock is rotated through various pastures to graze, thus controlling weeds and insects without chemicals.

In conclusion, intensive agriculture is practiced in the world without a break and has the following negative effects on soil: a. soil erosion; b. fertility impairment of soils; c. contamination of soils by pesticides, fertilizers and other biocides used during intense agricultural practices; d. salinization of soils; e. distancing from optimal agrotechnics; f. acidification of soils taking into account the methodologies employed by human beings to mitigate intensive farming as a source of soil pollution methods like crop rotation, mixed cropping, strip cropping and cover cropping can be employed to mitigate the effects of intensive farming on soil pollution (Pimentel & Burgess, 2013).

The use of intensive farming methods has developed as a result of agricultural technological advancements and the need to improve crop yields. Unfortunately, there are unintended consequences caused by intensive farming, including soil pollution. Around 85% of all land used for farming is affected by soil pollution in some form, caused by agricultural chemicals such as pesticides, fertilizers, and antibiotics. Examples like this demonstrate the need for innovative solutions to mitigate the effects of intensive farming on soil pollution, in order to reduce the impact of agriculture on the land and on public health (Zhao et al., 2015). The long-term solution is to place more land into conservation and reduce the use of fertilizers, pesticides and other chemicals in crop production. Research on the dynamics of soil pollution by intensive farming suggests a grim prognosis for the world's valuable arable land. Strict legislation and best-practice standards can help mitigate this problem, but using the most efficient technologies, such as precision agriculture, is the only way to reverse our current track. The use of pesticides and fertilizers has made intensive Farming a major contributor to soil pollution. The use of pesticides in agriculture pervades our entire world, which can have detrimental effects on soil. Fertilizers, manure, and human waste also impact soil quality because they enable the accumulation of excess phosphorus and nitrogen in the soil and this leads to water pollution. Human intervention is urgently needed in order to contain the damage already done by intensive farming practices, changing farming practices and promoting sustainable development are two key actions that can be taken right away.

3.12 PETROCHEMICALS

Figure 3.13: An overview of petrochemicals.

Source: Sui X, Wang X, Li Y, Ji H. Remediation of Petroleum-Contaminated Soils with Microbial and Microbial Combined Methods: Advances, Mechanisms, and Challenges. Sustainability. 2021; 13(16):9267. https://doi.org/10.3390/su13169267

Petrochemicals such as artificial pesticides, fertilizers and herbicides that are used for pest control, pollute the soil and can ultimately lead to food supply contamination. Petrochemicals can range from any substance derived from crude oil through the process of refining, especially, as a product of the chemically-altering refining of petroleum. Common products include gasoline, and kerosene derivatives such as diesel fuel and jet fuel, in general, the lubricants and greases and other chemicals (e.g., perfumes) made from them; plastics (e.g., polyvinyl chloride); pharmaceuticals; dyes; synthetic rubber; detergents; pesticides/herbicides/fungicides; explosives, solvents and fuels (Nearing, Lane, & Lopes, 2017). Generally speaking, petrochemicals are organic compounds that are isolated or produced by the petrochemical industry. Petroleum is made up of hydrocarbons (organics with only carbon/hydrogen bonds). Almost all petroleum contains some extra elements trapped within the hydrocarbons during formation (such as sulfur and nitrogen), either by virtue of being dissolved or chemically bonded or depending on the individual product being extracted these impurities may be undesirable. The hydrogen that petrochemicals add to organic molecules can be used to create other organic molecules. A refinery or chemical plant may convert ethane into ethylene by heating it to 700 °F.

Petrochemicals have been the main cause of soil pollution for about 100 years, where synthetic agricultural chemicals such as fertilizers, pesticides and herbicides are used in very large quantities creating runoff into rivers and lakes. Petrochemicals are potential hazards to the environment such as groundwater and drinking water exceeding their maximum concentration limits. Petrochemicals are potentially dangerous because of their toxicity, flammability, and explosiveness. The composition of the products from the incomplete combustion of fossil fuels depends heavily on the properties of the fossil fuels themselves and on the conditions under which combustion is carried out. For example, insufficient oxygen for complete combustion will produce carbon monoxide rather than carbon dioxide.

Petrochemicals are the basis for a very wide variety of petrochemicals, all of which can be used for various purposes. A special class of petrochemicals are used in the manufacture of lubricants, many plastic products and synthetic fibers. Every year, more than 1 million tons of these petrochemicals are handled in Germany. The causes of soil pollution by petrol or petrol products range from leakage of fuel tanks and fuel pipes over incorrect disposal to soil pollution during production or transport (Beilicci, Beilicci &Stefănescu, 2014). Petrochemicals are a group of environmental pollutants that are extremely toxic. The use of petrochemicals as fertilizer and pesticides resulted in serious soil pollution, which has some adverse effects on the crop quality and productivity. Soil pollution is the contamination of soil with man-made chemicals and takes many forms.

Many of these petrochemical products are emitted into the atmosphere or introduced into waterways or the soil. When they enter areas where wildlife and humans spend time, the chemicals may have harmful health effects. Petrochemicals can enter soil in a number of ways, including through leaks in production or transportation facilities. Another cause of petrochemical soil pollution is the improper disposal of hazardous materials such as used motor oil or paint. A major contributor to petrochemical soil pollution is the use of petroleum-based fertilizers. These fertilizers often contain harmful chemicals that add to soil pollution. Petrochemicals also seep into the ground whenever gas, oil and diesel fuels are spilled on our roads.

Chemicals used in oil extraction, refining and distribution are known as petroleum hydrocarbons (PHCs) or petrochemicals. In fact, this is the often-cited cause of pollution in the Niger Delta. PHCs have been widely used in Nigeria since the 1950s and in the Western Niger Delta since the 1970s when the first industrial-scale oil developments took place. Drilling, pipelines, storage tanks and other means of transportation can result in leaks and spills which may have a direct effect on soils. PHC compounds include gasoline, diesel and kerosene, as well as a range of chemical products from pesticides to detergents. These compounds are made up of a wide variety of chemicals that by themselves can cause harm to soil systems (Nigussie et al., 2015).

Environmental pollution and its subsequent effect on public health is a hotly debated topic. While water and air pollution often get more media attention, the quality of our soil is equally important. Because petrochemicals are one of the main causes of soil pollution, we need to mitigate their influence on our land as soon as possible. Although stringent regulations are necessary, they are not enough, because some of the companies that use petrochemicals have already proven they will violate them, putting US citizens in harm's way. One proposal is to heavily fine any company that uses petrochemicals and contaminates our soil. More so, a better understanding of how petrochemicals contribute to soil pollution is vital in designing potential solutions.

Scientists agree that it's essential to control the flow of petrochemicals into our land. What's more, soils in tropical rainforests are expected to shed a third of their carbon content by 2050, due to the high acidity caused by petrochemicals, according to a study published in Nature (2014). While this is largely theoretical, it suggests that soil acidity caused by fossil fuel use may be harmful to plants and animals, as well as humans. We have many solutions to this problem. Scientists have been exploring ways to transform petrochemicals into different substances that could be less toxic. We could also reduce the amount of petrol-based products that we use, but that would require an increase in public awareness (Molla &Sisheber, 2017).

There is evidence that the use of fertilizers containing organophosphate chemicals in household gardens may contribute to soil organic matter reduction and soil pollution, resulting in damage to plants and other life forms. The process of soil creation is slow, fragile and difficult to reverse. By choosing less toxic modes of fertilizers, households can help mitigate an important but often ignored source of pollution.

3.13 ROAD DEBRIS

We're all familiar with the sight of a bag of trash now and then blowing across an otherwise serene desert. Road debris can come from anywhere and it is often unavoidable, in fact, road debris is one of the leading causes of soil pollution. The obvious culprits are things like car wrecks (broken glass, oil, & rusted metal), littering (cigar butts, cigarette wrappers, & cups) & natural disasters (earthquakes, landslides &tornadoes). These are a few of the potential dangers to our soil health if these problems aren't addressed immediately.

First, the random nature of falling or blowing road debris means that the contaminant will not be evenly distributed. This can lead to localized problems around the site where the road debris has come to rest. Areas of vegetation may experience stunted growth, or areas of non-vegetation may become hazardous and inaccessible to people and animals because of toxins. Second, if the contaminants contained in road debris are harmful chemicals, heavy metals or other toxic substances, they can concentrate in the topsoil until they reach toxic levels. Once certain ground-dwelling organisms have absorbed those contaminants, they will become toxic too and be consumed by diverse wildlife. The toxins will work their way up the food chain directly affecting local bird and mammal populations and so on (Moges& Holden, 2007).

Road debris causes soil pollution by using sludge or chemical runoff, or salt water soaking into the ground and poisoning it. This causes erosion of the soil, loses topsoil down to deeper layers, and also contaminants that are leaching off of the roads. "Debris on roadways has potentially dangerous effects on local color and ecology", according to a report issued by the University of East Anglia in March last year. Road pollution plays an unknown role in climate change and could encourage occurrence of invasive species at the edges of roadsides. In mountainous areas, debris from roads is washed into streams where it harms ecologically important birds and bats and degrades water quality in other ways as well.

Road debris, such as oil and solvents, that has spilled, leaked or been dumped onto roads can wash into storm drains and then into our rivers and oceans. Not only can road debris be unsightly, it also is harmful to aquatic organisms such as fish and birds, if they swallow it. Road debris can also accumulate in the sediment of a water body and smother plant and animal life. Muck & debris from roads, streets or even parking lots can be a major source of contamination (Mekuria et al., 2009). Oil & grease is especially problematic because it contaminates soil and nearby water sources, and the roots of surrounding vegetation. Oil can be removed in the lab but the pollution will remain in place permanently unless remediation is performed by a licensed provider.

Figure 3.14: A sample image of road debris.

Source: Image by Wikiwand

A common misconception is that tire shops dispose of old tires by burying them in landfills. Actually, only a handful of states allow tire burial while others require the tires to be recycled. Tire recycling has significantly improved the way we handle and handle tires to convert them into a number of useful products, including road debris. This advantage has rather proved dangerous to our soils (Van Straalen, 2002). Scientific research states that many types of hazardous materials are used on the road and they often end up in the soil. These pollutants in turn cause a variety of problems for animals, plants, and humans as well.

Road debris, in the form of shredded tires, is a major contributor to soil pollution. It can introduce substances such as polycyclic aromatic hydrocarbons and heavy metals into the soil, affecting its biological integrity and altering human interaction with the soil. The current model of tire manufacturing is, in part, to blame for this degradation of our soil's health. The use of vulcanized tires can lessen breakage during operation while also decreasing road noise and improving fuel efficiency. By replacing existing tires with vulcanized models, we can mitigate road debris as a cause of soil pollution (Meshesha et al., 2012).

It is unethical to leave litter on the ground knowing that it will pollute our environment and harm our wildlife. This is especially true for harmful items like cigarette butts, which account for one-third of all roadside litter. An increase in the amount of road debris will certainly have an adverse impact on farmlands. This is because some of these wastes may get into the soil, thus causing pollution. Awareness of the pollution caused by road debris is increasing. In most cases, plastic and paper thrown out along roadsides can be found in nearby fields and streams. When we understand the negative impacts of this litter, we all want to prevent it.

Indeed, preventing this from occurring is a matter of proper waste disposal management. Along roadways and at construction sites, it is crucial that waste be stored appropriately to prevent unnecessary infiltration into soil and water systems. There are many different ways to mitigate road debris as a cause of soil pollution, but the two most efficient seem to be enacting laws against littering and having trucks cover their loads while traveling on roadways. It may be difficult to enforce the law, but if people know there is a law in place then they are more likely to think twice about littering out of fear of being caught (Ma, Zhao, & Zhu, 2021). Putting safety measures in place to ensure that truckloads are covered when traveling on roadways would prevent large amounts of gravel and soil from being accidentally spread along highways. This will not entirely eliminate such occurrences because there will inevitably be failure for some reason, but it will certainly reduce how much of this type of pollution is caused by cars and trucks.

As we see so often in the media - road construction projects, especially those that require blasting and drilling, are notorious for kicking up dirt and dust into the surrounding environment, polluting surrounding bodies of water and waterways, not to mention our lungs. The good news is we don't have to wait for the government or other large agencies to intervene. Together we can voluntarily become more responsible and learn ways of mitigating road debris, starting with learning how to dispose of our trash responsibly and recycle whenever possible.

3.14 DRAINAGE OF CONTAMINATED SURFACE WATER INTO THE SOIL

Drainage of contaminated surface water into the soil as a cause of soil pollution is a key component in modern farms, especially with the use of gravity feed water systems to move chemicals around. Drainage of contaminated surface water into the soil is one of the common causes of pollution in the soil. The pollutants are usually dissolved nutrients and fertilizers like nitrogen, phosphorus and dissolved metals that can cause environmental and health problems. The quality of irrigation water is a key factor in assessing the economic, environmental and health risks associated with soil pollution.

Soil pollution by heavy metals can be carried out by contaminated surface water, the main source of this type of pollution is industrial and agricultural waste. In surface water runoff, many pollutants are carried by the stream and eventually deposited into the soil where they pose a threat to human health. These pollutants can reach the groundwater supplies and contaminate the water source (Lim et al., 2020). Soil is a three-dimensional material that contains solid, liquid and gaseous phases with varying degrees of interconnectivity. Soil pollution is the presence of contaminants in the soil that may cause an adverse effect on human health or the environment. Contaminants released into the natural environment have severe economic and ecological implications. The most common sources of contamination by humans are urban and industrial activities that generate toxic pollutants.

According to the United States Geological Survey, surface water pollution is caused by natural or human sources contaminating water bodies. The survey has identified a variety of contamination are sources that can infiltrate or run off into surface water, such as pesticides and herbicides from commercial agricultural operations; pathogens generated by improperly disposed of waste; improper management of animal manure; oil and other contaminants from urban storm-water runoff; and salt accumulation from agricultural irrigation.

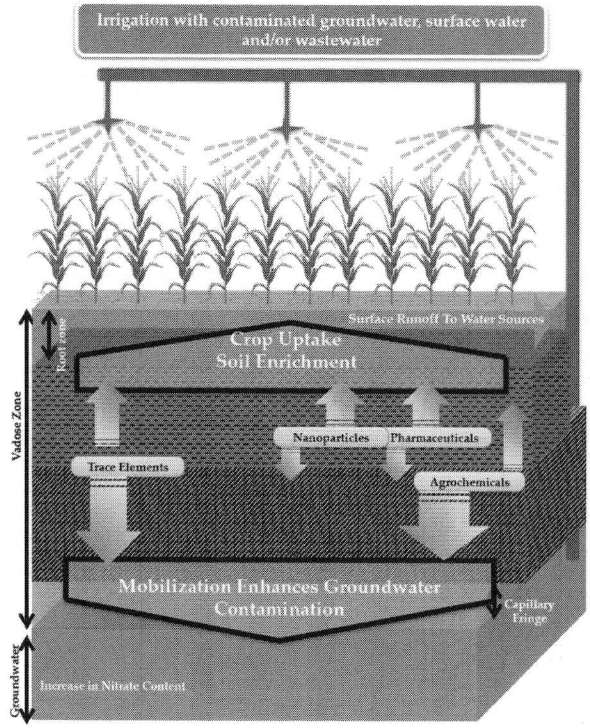

Figure 3.15: Contaminated surface water can run and eventually affect the soil.

Source: Malakar A, Snow DD, Ray C. Irrigation Water Quality—A Contemporary Perspective. Water. 2019; 11(7):1482. https://doi.org/10.3390/w11071482

Drainage of contaminated surface water into the soil can indicate a problem for the rest of the community. The Water Act imposes legal obligations on farmers for the drainage of contaminated water onto the land. When contaminated water enters the soil, it is possible that dangerous substances are taken up by plants, which may then form a health hazard to people. To help ensure that dangerous waste is not drained into the ground, you must take steps to determine whether the water contains any pollutants. Soil pollution occurs mostly by soil contamination via man-made chemicals, including a variety of contaminants ranging from pesticides to heavy metals and toxic organic compounds. Over 95% of such pollution is caused by irrigation drainage water containing fertilizers, human sewage and pesticides. The five most common groups of pollutants are hydrocarbons, hazardous substances, heavy metals, radioactive substances and urban wastes (Li et al., 2018).

The process of drainage from areas of higher elevation to those at a lower elevation, known as surface runoff, is a primary cause of soil erosion. A common way that humans contribute to soil erosion is by using the land improperly for agricultural uses such as tilling fields that are steeper than they should be. When these fields are cultivated, the soil will be affected by water drainage through the field to a lower elevation because of gravity. This can result in serious lossesss of topsoil and lead to long-term damage to the land's ability to support agriculture or development. Dump leaks at industrial or nuclear waste sites cause contamination of soil and groundwater. Clean up is expensive, and often inadequate. Site identification through remote sensing is necessary to prevent contamination cross-contamination.

Contaminated surface water can also result from man-made substances that are deliberately or accidentally introduced into the environment. Large amounts of such chemicals can build up in soil, air and water over a period of years, and can be highly toxic to animals, plants and the people who depend on these for food (Blake et al., 2021). Mitigating contaminated surface water from draining onto the ground is an important and complex goal to keep our soil clean. One can prevent runoff of concentrated materials by fencing off your storage areas and processing areas so that rain runoff and draining water cannot reach the soil. And, use a concrete foundation to keep these materials out of contact with the soil.

A whole lot of individuals want to mitigate the drainage of contaminated surface water into the soil as a cause of soil pollution. Some studies demonstrate that mitigating drainage of contaminated surface water into the soil as a cause of soil pollution can diminish reduced mental health impacts and behavior problems associated with exposure to chemicals, thereby augmenting our happiness. Mitigating drainage of contaminated surface water into the soil as a cause of soil pollution means cleaning up the surface water and eliminating chemical exposure to the soil. The most effective way to prevent contamination of surface water is to eliminate exposure by cleaning up the surface water. If a natural disaster has caused a chemical exposure to soil, then this practice is critical in minimizing damage to people's health.

3.15 WASTE DISPOSAL

Oil and Fuel Dumping;

Oil and fuel dumping is a leading cause of soil pollution. The United States Environmental Protection Agency (EPA) has defined soil pollution as an "a change in the natural character of soils resulting from human activity." The EPA has also stated that petroleum-based contamination (from oil, gas, or diesel products) can travel through porous soils and contaminate groundwater supplies that are potentially used for watering crops and livestock as well as drinking water for people.

Oil dumped into the ground can cause soil pollution. The oil will sink through to the groundwater, which can become a source of drinking water for humans and animals as well as plants. The Health & Safety Executive estimates that there are 400-500,000 instances of oil contamination in the UK every year. Some of this comes from spills when filling your car and some from deliberate dumping of waste. Aircraft will discharge oil and fuel while flying (Blake et al., 2018). The crude fuels may enter the soil and cause serious problems if dumped inappropriately, like near an ecosystem with small plants and animals in it. A major challenge of the 21st century is to learn how to manage waste while minimizing adverse effects on human health and the environment. Oil dumping threatens the environment as oil can penetrate into the soil, destroying vegetation and poisoning wildlife, and if it gets washed by rain into water sources, it can cause the death of aquatic life.

Nuclear Wastes;

Nuclear waste may cause soil pollution and disrupt the sustainable ecosystem of human beings. With increasing nuclear waste disposal, soil pollution will become a time bomb, which may seriously endanger human safety when it goes off. It is likely that, in recent years, nuclear accidents have had a profound impact on the environment. Specifically, examples include harmful mining byproducts, chemical spills and oil spills in landfills. Nuclear waste disposal may be one of the main concerns of mankind. First and foremost, it is believed that the effects of nuclear waste disposal are the cause of many soil pollution cases around the world. A controversial issue merits serious attention based on some of its implications.

Elevated levels of toxic elements have been found in soil adjacent to nuclear waste sites and underground repositories constructed in fractured rock formations. Nuclear waste disposal is the act of containing hazardous material in a location where the material will be isolated from the environment for an extended period of time. The major cause of soil pollution is human activities (Lal, 2019). Soil pollution caused by nuclear waste stems from the way in which it is disposed of. Nuclear waste can only be stored for a limited amount of time in its storage containers before needing to be put into a repository that is designed for long-term storage, also described as permanent storage. These facilities are constructed deep underground and prevent the radiation from reaching the surface level. In fact, all of the radioactive material has been buried underground for millions of years.

Direct Discharge of Industrial Wastes to the Soil

Industrialization in an agricultural country cannot be avoided. Meanwhile, it causes the problem of soil pollution due to the direct discharge of industrial wastes into the soil. The problem is caused by the ignorance of the farmers when they use chemicals as fertilizers and pesticides. The chemical fertilizers which are used directly, whether it's productive or not, will stay and be attached to soil. Those are dangerous products that can cause root poisoning eventually killing the plant. The same thing happens when you spray pesticides either organophosphates or carbametes. These substances will be very hard to degrade and will become the food chain for human consumption.

Industrial wastes which contain heavy metals, cyanides and other deleterious substances pollute the soil. This can cause a decrease in the usefulness of soil for growing crops, as well as damage to agricultural productivity. The problem is not confined in developed countries, but also challenging in developing countries due to weak regulatory mechanisms. Agrochemical residues, pesticides and petrochemicals are the major forms of soil pollution that are already contributing to the poisoning of natural food chains. In order to better understand how industrial wastes contaminate the soil, a literature review was conducted (Lefèvre et al., 2021). The review shows that industries often use contaminated water for cooling purposes and land surface runoff discharges them into the nearby rivers or lakes that ultimately reach the soil after heavy rainfalls. Soil pollutions have adverse impacts on the agricultural sector and it becomes an urgent need to address this issue through sustainable development strategies. Soil conservation practices, wastewater recycling, reduction in pesticide use and strict implementation of environmental laws can improve the situation.

Discharge of Sewage;

The direct discharge of sewage into the soil has become a cause of environmental pollution. If a large amount of organic substances are added to the soil, they will be decomposed by microorganisms in the soil, which will consume the oxygen in the soil and render it devoid of oxygen. In this case, not only denitrification is promoted, but also many harmful substances are generated by anaerobic fermentation. Therefore, pollutants can transfer from soil water to plants through various ways such as root absorption, stem cutting, canopy infiltration and so on. The transportation of pollutants along the food chain can also occur (Chuchkalov, Fadeev, & Alekseev, 2020).

Direct discharge of sewage to the ground, accelerated the death of soil organisms, and also promoted the growth of some microorganisms, but will affect the normal growth and reproduction, so microbial activity declines. Direct discharge of sewage into the soil is beneficial to a small part of plants, while having a strong inhibitory effect on other plants. There are direct discharge toilet wastewater containing live bacteria such as *Escherichia coli* , etc., if it is directly discharged into the soil, will affect the quality of agricultural products.

Landfill and Illegal Dumping

Landfill and illegal dumping are really a big problem that we are facing nowadays. There have been many cases where people have dumped illegal trash in one area, and the consequences of this problem have been something that has affected the environment for many years. One thing that we can do to solve this problem is to educate the people about the environmental impact of improper waste disposal, and also to provide them with proper ways of doing so (Kowalska et al., 2018).

Landfills and materials dumped illegally contribute to soil pollution. The trash in landfills and materials dumped illegally eventually decompose or leach their chemical compounds into the topsoil. This process is often assisted by scavengers and other animals that wander into landfills looking for food in the waste. Once these chemicals enter the ground, they become a part of the soil, polluting it for generations to come. Solid wastes can be disposed in landfills, or by illegal dumping. Both methods endanger soil fertility and soil contamination and pose serious risks of groundwater contamination. The majority of solid waste disposal occurs in landfills, which are among the most problematic methods of solid waste disposal.

Coal Ash

Coal ash, waste matter produced by the combustion of coal, can cause serious soil pollution. Coal represents the single largest source of electricity in the world and has helped thousands move out of poverty. But this is also why the industries presents a significant environmental risk, as their operations leave behind a trail of dangerous byproducts. Coal ash (a particularly hazardous byproduct) causes soil pollution, pollutes waterways and can easily be inhaled. Soil pollution caused by waste ash is a serious problem. Coal-burning power plants tend to be located far from the main consumer markets and are usually therefore sited in remote areas. Spills, leaks and drainage from these power plants pose major threats to the environment (Keesstra et al., 2018).

The fly ash from coal-fired power plants can affect soil quality and the environment. Airborne coal ash could be brought into our home, destroy the carpet and make more dust. We need to use a dehumidifier as our home has a mildew problem,. Soil and sediments on site have been contaminated with arsenic and manganese. Once it enters our water supply it can cause health problems. People who live within the surrounding areas have an increased risk of cancer and heart disease.

Electronic Waste;

Electronic waste is becoming a growing problem as more electronics are placing and creating landfills. Electronic waste is becoming a cause of soil pollution and is destroying the ozone layer in the atmosphere. Electronic waste is the common term for electronic products that are nearing the end of their 'useful life. This can be just about any device with circuitry or electrical components. If not properly disposed of, electronic waste can lead to soil pollution as well as water and air pollution. When different chemical constituents of electronic devices seep into the soil, they contaminate it and have adverse effects on both humans and animals.

Rocks Containing Large Amounts of Toxic Elements

Rocks containing large amounts of toxic elements such as calcium sulfate, lead, magnesium and potassium are the cause of soil pollution. Most soils contain some other elements in larger or smaller amounts. They become sources of soil pollution when they are broken down by weathering and erosion into tiny particles. When these particles are deposited in the soil, plants take up the toxic elements and pass them on to animals that feed on the plants. In addition, toxic chemicals dissolve from the rocks into ground water, which is a source of drinking water for people - this may also include fish and other aquatic animals (Kelly et al., 2020).

It goes without saying that the presence of toxic elements in the soil will have a detrimental effect on the health of both plants and animals. So, naturally, a popular place to dump toxic wastes is often remote regions where they will come into contact with rocks and soil. The problem is that various substances can seep through rock and contaminate the soil, thereby posing an equally serious threat to plant life. The formation of metal-contaminated soils is a slow process. It begins with the weathering of metal-rich rocks, followed by the development of soils and their short-term biosphere. Thus, the concentration of elements that may be toxic varies depending on several factors: the nature of the parent rock, the time required for weathering, the biological activity and chemical reactions. As a result, chemical anomalies are created in patches of soil that are called hot spots.

Rocks are the ultimate source of metal ions that are present in the soil and water. Agricultural activities (irrigation) influence metal fate and transport in continental surface waters. Metal pollution has become a major environmental issue worldwide, mainly due to its presence in agricultural soils and its subsequent transfer to the human food chain. There is no doubt that releasing large amounts of chemical substances, heavy metals or nutrients into the soil, groundwater system and surface water bodies from anthropogenic sources causes pollution

Contamination by Lead due to Vehicle Exhaust, Cadmium and Zinc Caused by Tire Wear.

High levels of lead, cadmium and zinc in urban soils are a result of contamination by vehicle exhaust and rubber tire wear. This is the first time that studies have shown that the pollutants of lead (Pb), cadmium (Cd) and zinc (Zn) in the exhaust gas of automobiles can be absorbed by tire rubber, fall onto road soil and cause soil pollution. This study shows that in order to reduce SO2 emissions, it is not enough to just develop new vehicles with low emissions (Kheyrodin, 2016). It also shows that in order to create a peaceful environment with clean air and clean water, a comprehensive management system must be set up for different types of polluting sources, including vehicle exhaust gases. Tire wear results in cadmium, zinc and lead being washed into the soil. Lead can cause nervous system damage in children, cadmium can be linked to kidney disease and zinc can result in metal poisoning.

The extremely high output of these harmful elements by car traffic and the consequent contamination of surface and groundwater is widely considered to be a major cause of the worldwide increase in lead, cadmium and zinc pollution in the last decades. This is why researchers are finding high levels of cadmium, lead and zinc in soil surrounding roadways. The particles from the tires contaminate the soil and are extremely bad for plants and other organisms whose lives depend on its cleanliness of the soil.

Contamination by Strengthening Air Pollutants by Incineration of Fossil Raw Materials

As the fossil resources become scarce, the incineration of fossil raw materials will be strengthened. It is necessary to examine the effects of such efforts on soil pollution. Air pollution will become a social problem in the future. Strengthening incineration of starches for electricity generation and digested sludge for landfill disposal are considered to have an effect on this problem. By incineration, heavy metal contained in fossil raw materials is naturalized into ash, burnt gas and dust. Ash and dust are returned to the soil through water, but the gas is emitted into the air. However, there is no effective method to break these molecules by direct emission into the air yet now (Kumarasinghe, 2021).

The overuse of fossil raw materials causes soil pollution to rise exponentially in the current global system, a fact that can only be abated in the long term by a relentless search for further sources and progressive recycling. An immediate end to the overuse of raw materials is necessary.

CHAPTER 4

CONSEQUENCES OF SOIL POLLUTION

CONTENTS

4.1 What Is Soil Pollution? 78

4.2 Soil Pollution Causes and Types 78

4.3 Soil Pollution Consequences 80

4.4 Significant Impacts of Soil Pollution According to IPBES and the Fao 80

4.5 Damage to Health 81

4.6 Poor Harvest 87

4.7 Climate Change 90

4.8 Water and Air Pollution 93

4.9 Population Displacement 96

4.10 Soil Contamination and Species Extinction 99

4.11 Biodiversity Conservation and Poverty 101

4.12 Economic Impact 102

4.1 WHAT IS SOIL POLLUTION?

Soil pollution is when the concentration of the pollutants on the surface builds up to the point where it impairs land biodiversity and jeopardizes human health, notably through food. Heavy metals and other natural and artificial chemical compounds damage the soil in the same manner that chemicals, herbicides, and fertilizers used in animal breeding and intensive farming do (Kavian et al., 2019).

Figure 4.1: Soil Pollution.

Source: By Nils Ally - Personal trip in 2010, CC BY 3.0, https://commons.wikimedia.org/w/index.php?curid=14730037

4.2 SOIL POLLUTION CAUSES AND TYPES

Modern soil degradation is primarily caused by erosion, loss of organic carbon, increased salt content, compacting, acidification, and chemical pollution. In addition, the FAO distinguishes between two categories of soil pollution:

When the concentration of contaminants on the surface reaches a level that threatens land biodiversity and jeopardizes human health, most notably through food, this invisible disease manifests. Heavy metals and other naturally occurring and synthetic chemical compounds degrade the soil in the same way as pesticides, herbicides, and fertilizers used in animal breeding and intensive agriculture degrade the soil (Cerda et al., 2018).

Pollution caused by particular sources, which occurs in small regions and has readily identifiable causes, is specific pollution. Pollution of this nature may be found in cities, abandoned industrial sites, highways, illicit dumping, and sewage treatment plants.

Pollution is a huge problem that affects a large area and has many origins that are difficult to pinpoint. Pollutants are distributed over air-ground-water systems under these conditions, posing significant health and environmental risks.

According to the FAO, industry, mining, military operations, garbage — including technological waste — and wastewater management, farming, animal breeding, and the construction of urban and transportation infrastructure are among the primary sources of soil pollution induced by human activity.

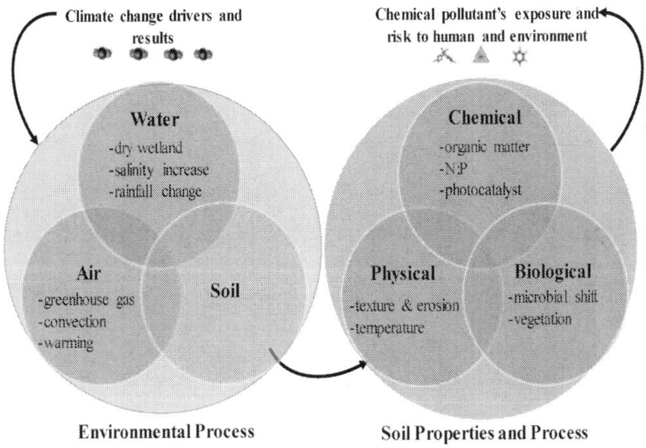

Figure 4.2: Ecosystem impairment caused by soil pollution.

Source: Biswas B, Qi F, Biswas JK, Wijayawardena A, Khan MAI, Naidu R. The Fate of Chemical Pollutants with Soil Properties and Processes in the Climate Change Paradigm—A Review. Soil Systems. 2018; 2(3):51. https://doi.org/10.3390/soilsystems2030051

4.3 SOIL POLLUTION CONSEQUENCES

For this topic, we will be looking deeply into the consequences of soil pollution.

Figure 4.3: What Are The Harmful Effects Of Soil Pollution?

Source: Amin Al Manmi DAM, Abdullah TO, Al-Jaf PM, Al-Ansari N. Soil and Groundwater Pollution Assessment and Delineation of Intensity Risk Map in Sulaymaniyah City, NE of Iraq. Water. 2019; 11(10):2158. https://doi.org/10.3390/w11102158

Toxic compounds accumulated on the earth's surface endanger human health and the quality of our food, water, and air.

4.4 SIGNIFICANT IMPACTS OF SOIL POLLUTION ACCORDING TO IPBES AND THE FAO

I. Damage to health

Pollutants from the soil enter our bodies via the food chain, producing sickness. Furthermore, the widespread use of antibiotics raises bacterial resistance to these drugs.

II. Poorer harvests

Soil pollution agents jeopardize global food security by lowering crop quantity and quality.

III. Climate change

Soil degradation contributed between 3.6 and 4.4 billion tons of CO2 into the atmosphere in the twenty-first century's first decade (Jomaa et al., 2013).

IV. Water and air pollution

Soil degradation impacts air and water quality, especially in emerging nations.

V. Population displacement

By 2050, soil degradation and climate change will have displaced between 50 and 700 million people.

VI. Species extinction

Soil pollution is one of the critical causes of the world's sixth mass extinction, with the number of terrestrial vertebrates dropping by 38% between 1970 and 2012.

VII. Desertification

In 2050, the world's driest regions may account for 45 percent of the global population, while global wetland areas have decreased by 87 percent during the last three centuries.

VIII. Economic impact

The global economy is predicted to lose more than 10% of its yearly GDP due to soil deterioration (GDP).

4.5 DAMAGE TO HEALTH

The consequences of soil pollution on human health have been extensively researched, particularly the effects of heavy metals in metropolitan areas, mining areas, nearby industrial sectors, and combat-damaged places. Such study has also been conducted in agricultural regions (Cogo, 2022).

Historically, heavy metals have been the focus of most research on the influence of soil chemistry on human health. To assess the degree of soil contamination and its potential impact on human health, indices such as the contamination factor, accumulation index, enrichment factor, contamination degree, sum of pollution index, single pollution index, ecological risk index, integrated pollution index, Nemerow pollution index, pollution load index, hazard index, dermal absorption factor, and aggregated carcinogenic risks have been developed. These markers help determine the extent of soil contamination and the risk of human exposure. For further information on these indices, please check available literature. While certain metals (such as copper, iron, and zinc) are necessary for plant development, their overabundance can harm plants and expose humans to disease risks. Heavy metal levels can damage the blood, liver, brain, kidneys, lungs, and other organs. Chronic exposure to heavy metals, even at low concentrations, can cause neurological and physical deterioration, as well as cancer (e.g., Parkinson's disease and Alzheimer's disease). In recent research, the effects of excessive heavy metal concentrations on human health have been widely examined.

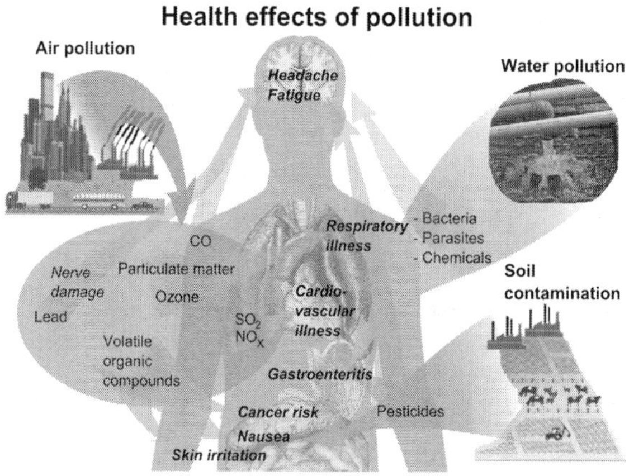

Figure 4.4: Health effects of pollution.

Source: https://aida-americas.org/en/blog/air-pollution-latin-america-and-its-effect-our-health-and-climate

Other soil chemistry studies looked at the effects of persistent organic pollutants (POPs) and radionuclides on human health. Grindler identified a significant link between POPs levels in urine and menopause in its early stages in the United States of America. Polychlorinated biphenyls impact prenatal development and birth weight (PCBs). Dichlorodiphenyltrichloroethane (DDT), dichlorodiphenyldichloroethylene (DDE), and hexachlorobenzene (HCB) are other POPs that affect birth weight (HCB). POPs are another name for endocrine disruptors. POPs build up in soils due to pesticide use and intentional release into the environment. Nonetheless, non-pesticide POPs have been intentionally and unintentionally released into the atmosphere.

Pesticides used in agricultural areas have been linked to an increased risk of chronic illnesses, including diabetes, cancer, and asthma, as well as a variety of short-term adverse effects like dizziness, nausea, skin and eye irritation, and headaches (e.g., dizziness, nausea, skin and eye irritation, and headaches). Each year, pesticide poisoning is projected to affect 25 million agricultural workers. Animal pesticides are not the primary cause of health risks. Herbicides like glyphosate have also been linked to adverse health effects on people (Issaka& Ashraf, 2017).

Cancer and leukemia are linked to radionuclides, which occur naturally in soil or are brought in by human activity (e.g., medical and nuclear waste) (e.g., medical and nuclear waste). The Chernobyl and, more recently, Fukushima disasters demonstrated the dangers of radionuclides on human health. Contamination of agricultural commodities was linked to radioisotope levels in soils following the Chernobyl disaster by consumption of crops and livestock grown on contaminated soils. According to an assessment in a Chernobyl-affected area by Komissarova and Paramonova, multiple locations continued to violate the 137Cs safety guideline by 3.5 to 6 times in 2017. Despite its high concentration, the radionuclide had minimal effect on crops and forages. Following the Fukushima disaster, it was expected that Japanese soils would be heavily contaminated with 137Cs, posing a long-term risk to agricultural commodities and human health.

Chemicals used in combat also contribute to soil pollution. Other harmful substances, heavy metals, and radionuclides are discharged into the soil, including energetic compounds, nitro aromatic explosives, organ phosphorus nerve agents, and petroleum products. Grounds have been shown to contain the most harmful contaminants linked with conflict. These include energetic compounds, nitro aromatic explosives, and nerve agents, including organ phosphorus.

Inhalation, skin absorption or penetration, and ingestion expose humans to soil chemicals, pathogens, and minerals. Numerous indices have been developed to measure the risk of cancer caused by ingesting, inhalation, or skin contact with contaminated soil. Wind-borne dust causes the inhalation of soil particles. Pneumopulmonary fibrosis, chronic obstructive pulmonary disease, sarcoidosis, and asthma have all been linked to dust movement (Ighodaro, Lategan, & Yusuf, 2013). This is especially common in persons who have weakened or damaged immune systems, such as children, the elderly, and those who already have chronic cardiac problems. Chemicals, such as arsenic, have been proven to be conveyed in dust particles, posing a threat to human health. Dust storms are widespread in arid and semi-arid environments.

On the other hand, human activity is increasing the number of dust occurrences in temperate continental, tropical monsoon, subtropical monsoon, continental monsoon, and temperate oceanic climates (e.g., road expansion, urban sprawl, agriculture, and mining). Previous research has found that skin contact with soils with high levels of heavy metals can cause pain and rashes. Heavy metal levels in the blood can be raised by consuming contaminated soil particles. Plants can absorb a wide range of heavy metals and swiftly transport them up the food chain. This issue has been noticed in wastewater-irrigated areas and farms and gardens near the urban, industrial, and mining sectors. While the effects of heavy metals on human health are evident via skin contact, the dangers of inhalation and ingestion are significantly greater.

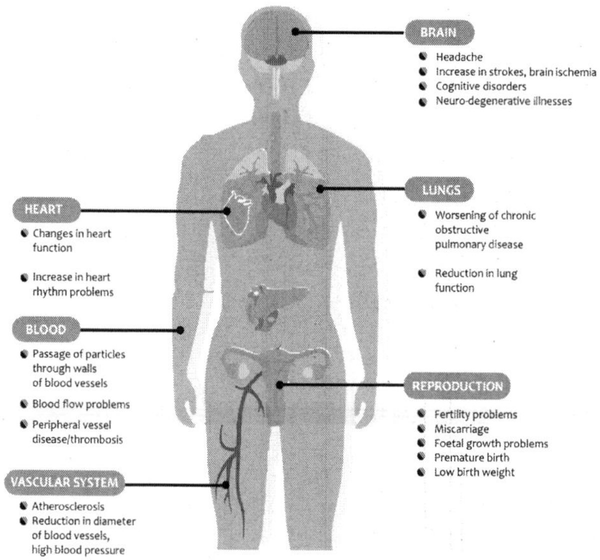

Figure 4.5: Soil pollution and risk to human health.

Source: Ali, M. M. , Hossain, D., Al-Imran, A., Khan, M. S. , Begum, M., & Osman, M. H. (2021). Environmental Pollution with Heavy Metals: A Public Health Concern. In M. K. Nazal, & H. Zhao (Eds.), Heavy Metals - Their Environmental Impacts and Mitigation. IntechOpen. https://doi.org/10.5772/intechopen.96805

Urban agriculture is increasingly being promoted to improve the quality and quantity of soil-related ecosystem services. Each year, urban agriculture is expected to generate ecological benefits worth between $80 and $160 billion worldwide. Furthermore, urban agriculture supports food security, poverty reduction, community development, and social justice; it reduces energy consumption and carbon emissions; it increases the resilience of urban people; it creates employment and stimulates the local economy and education (Haregeweyn et al., 2015). Each of these has the potential to improve human health. Urban agriculture has the potential to make a substantial contribution to sustainable development and the achievement of the United Nations Sustainable Development Goals (for example, poverty eradication, hunger elimination, and sustainable cities and communities).

Numerous papers on the mapping of soil heavy metal pollution in urban, industrial, and agricultural regions and at the continental scale have been published. Lately, heavy metals in soils have been mapped using modern methods, including proximal and distant sensing. These procedures represent a significant advancement in mapping, allowing for processing a higher number of sample points in less time than more traditional methods. Understanding pollution sources, variables influencing soil contamination distribution, and the population exposed to soil contamination all necessitate spatial patterns. As a result, comprehensive mapping is necessary to avoid the adverse health effects of soil pollution. Numerous maps have been created to depict soil radionuclides in Chernobyl-impacted areas and track human exposure to linked radionuclides. In the aftermath of the Fukushima disaster, Japan employed a similar strategy. Numerous studies have found a relationship between heavy metal mapping and radiation. Future soil degradation and human health requirements

As previously stated, the link between soil pollution from many sources and human health is well documented. Heavy metals, radionuclides, and chemical compounds are all examples of this. Recognizing that these difficulties are a concern in modern society, it is necessary to look ahead and predict how these challenges may spread in the future, including probable amplification.

Toxins travel long distances in the dust. Dust storms are expected to become increasingly common due to climate and land-use change (e.g., urbanization, agricultural intensification, and desertification). As a result, it is reasonable to expect a rise in diseases linked to climate change. Some attempt has been made to identify the origins and type of content being supplied and link this to human health. However, more research is needed to predict these events and find strategies to mitigate their influence on human health, such as natural remedies. The majority of study has focused on heavy metal contamination (Ha, Van Dung & Ngoc, 2013).

Nonetheless, research into plastics, pesticides, and similar chemical compounds has increased dramatically in recent years. The numbers of chemical components in soil that have been identified as dangerous to human health has risen significantly. Analyzing the geographical distribution of soil microplastics, pesticides, and other contaminants is critical for identifying areas with potentially hazardous soil conditions.

It is also necessary to determine the amounts of these components in soil that are judged harmful to human health. While most research has focused on specific contaminants, these pollutants interact with one another and the larger complex soil ecology. Methods for evaluating the toxicity of chemical combinations and their interactions in the soil must be developed.

Despite the importance of urban agriculture, multiple studies have found that food grown in urban areas may have higher levels of heavy metals. This is a severe issue that must be addressed to protect the health of those who consume food grown in cities. Brownfield gardening and wastewater irrigation can cause serious contamination issues with the crops grown in these places. Increased pollutant levels in vegetables are not harmful to human health in several metropolitan areas, including Barcelona, Spain; Sevilla, Spain; Lisbon, Portugal; Madrid, Spain; Sheffield, United Kingdom; and Braganca, Portugal. However, pollution levels in numerous places, including Rio de Janeiro, Brazil, Daejeon, South Korea, Rome, Italy, Melbourne, Australia, Dera Ismail Khan, Pakistan, and Ghaziabad, India, were high, posing a hazard to human health (Guerra et al., 2017). Heavy metal poisoning of fruits and vegetables created the most severe problems in India, Pakistan, and China. As a result, future studies on urban food production must extensively investigate both the good and negative consequences for human health.

Soil maps can help investigate the connection between soils and human health. The sampling technique used to collect data, the methodology utilized to project point data onto a two- or three-dimensional mapping output and the mapmakers' ability and training all contribute to the accuracy of our maps. As a result, more excellent study into successful soil mapping methodologies is necessary, including data collection, analysis, modification, and presentation.

4.6 POOR HARVEST

Figure 4.6 :(a) Aerial view of a typical erosion gully in the Cerrado of Brazil. On the right, soybean plantation showing bare soils after harvest, on the left, riparian vegetation surrounding a stream. In the foreground, an earth road and two erosion gullies of different ages. The erosion gullies breach the riparian vegetation (capture date of photographs July 1995); (b) Active erosion gully in Figure (a). Deep lateral piping hollows drain groundwater from the surroundings, which flows off through the channel even during the dry season; (c) Stream receiving sediment from the active erosion gully in Figure (b) during the dry season; (d) Pioneer vegetation developing on the bottom of the erosion gully in Figure 1c during the dry season. © Karl M. Wantzen.

Source: Wantzen KM, Mol JH. Soil Erosion from Agriculture and Mining: A Threat to Tropical Stream Ecosystems. Agriculture. 2013; 3(4):660-683. https://doi.org/10.3390/agriculture3040660

According to recent research, soil pollution poses a severe danger to agricultural output, food safety, and human health. It was announced on May 2 by the United Nations Food and Agriculture Organization (FAO). According to the study, industrialization, war, mining, and agricultural expansion have all had a significant impact on the world's soils. As cities expand, the ground is increasingly being exploited as a landfill for vast amounts of urban trash. "Soil pollution has an influence on the food we eat, the water we drink, the air we breathe, and the health of our ecosystems," according to an FAO press statement (García-Ruiz et al., 2017). According to the authors of "Soil Pollution: A Hidden Reality," far too little is known about the extent and severity of soil pollution, which they define as "the presence in the soil of a chemical or substance that is out of place and present at a higher than normal concentration, causing adverse effects on any untargeted organism." According to their examination of the current scientific literature, the bulk of study has been undertaken in developed economies, showing significant gaps in our knowledge of the problem's underlying nature.

The brochure includes facts and figures that depict worldwide soil stress. Soil pollution, for example, is predicted to affect 80,000 places in Australia. Polluted soils account for 16% of all soils in China and 19% of agricultural soils. Furthermore, the European Economic Area and the Western Balkans include around 3 million contaminated sites. The United States has 1,300 pollution hotspots, according to a national registry. Human activities is mostly to blame for soil pollution. According to the study, the principal anthropogenic causes of soil pollution include chemicals used in or byproducts of industrial operations, home, animal, and municipal wastes (including wastewater), agrochemicals, and petroleum-derived products. "These chemicals are delivered into the ecosystem either mistakenly or actively, as with oil spills or landfill leaching, or purposely, as with fertilizers and pesticides, untreated wastewater irrigation, or sewage sludge land application." However, soil contamination is induced by smelting, shipping, pesticide spray drift, and incomplete combustion of a variety of pollutants.

According to the authors of most books, the main causes of soil contamination include excessive fertilizer and manure application, as well as incorrect use of the major elements – nitrogen (N) and phosphorous (P) – in fertilizers: "Excessive fertilizer usage can cause soil salinity, heavy metal building, water eutrophication, and nitrate buildup." Furthermore, the fertilizer business is well-known for its heavy metal supply." Despite its potential agricultural value, bovine dung may include significant concentrations of heavy metals, pathogen organisms, and antibiotics, boosting the growth of antimicrobial-resistant bacteria in soils treated with such manure, according to the paper (Frankl et al., 2016). Global animal manure output rose by 66 percent between 1961 and 2016, from 73 to 124 million tonnes of nitrogen, with manure applied to soils growing by 18 to 28 million tonnes of nitrogen and N intake from pasture dung increasing by 48 to 86 million tonnes of nitrogen.

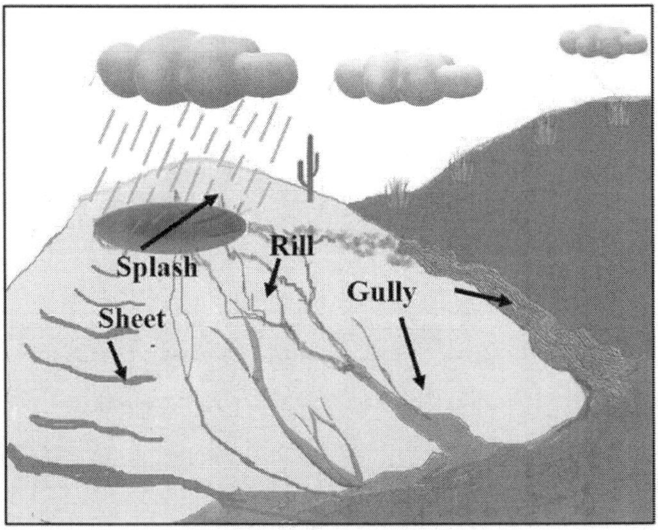

Figure 4.7: Soil Erosion – Causes and Effects.

Source: Senanayake S, Pradhan B, Huete A, Brennan J. A Review on Assessing and Mapping Soil Erosion Hazard Using Geo-Informatics Technology for Farming System Management. Remote Sensing. 2020; 12(24):4063. https://doi.org/10.3390/rs12244063

The usage of pesticides is another concern. According to the FAO, pesticide usage has grown in several low- and middle-income nations during the previous decade. Bangladesh, for example, has expanded its use by four-fold, Rwanda and Ethiopia by more than six-fold, and Sudan by ten-fold. Pesticides are misused when applied in more significant quantities than necessary, and when practices that contribute to their spread into the environment, such as spraying with inappropriate application equipment or planes over vast areas, harm people and non-target organisms, are employed, according to the authors. Soil pollution has a detrimental influence on food security since it reduces plant metabolism and hence agricultural productivity and makes crops hazardous to eat for animals and people. Pollutants also impact soil microbes and, more significantly, soil-dwelling species, diminishing biodiversity and fertility in the soil (Fenta et al., 2021).

"Soils' ability to cope with pollution is limited; preventing soil pollution should be a major worldwide priority," Maria Helena Semedo said. The FAO advises that national governments impose limitations on soil pollution and prevent the buildup of pollutants above predefined thresholds from preserving human well-being, a healthy environment, and safe food. According to the group, governments should also allow for the restoration of polluted soils that have exceeded permitted levels.

4.7 CLIMATE CHANGE

Soil degradation contributed between 3.6 and 4.4 billion tons of CO_2 into the atmosphere in the twenty-first century's first decade. Carbon dioxide levels have reached unprecedented heights in at least 800,000 years. In reality, CO_2 levels in the atmosphere were this high over 3 million years ago, during the Mid-Pliocene Warm Period, when temperatures were 2°–3°C (3.6°–5.4°F) higher, and sea levels were 15–25 meters (50–80 feet) higher.

The usage of fossil fuels for energy is causing carbon dioxide levels to rise. Plants extracted carbon from the atmosphere over millions of years through photosynthesis; people returned that carbon to the environment in just a few hundred years.

Figure 4.8: Land degradation worsening climate change- serious land degradation in Nauru after the depletion of the phosphate cover through mining.

Source: Public Domain, https://commons.wikimedia.org/w/index.php?curid=757619

Humans exhale more carbon dioxide into the atmosphere every year than natural processes can remove, resulting in a net increase in world carbon dioxide levels. The faster we advance year after year, the more we outgrow what biological processes can remove. In the 1960s, carbon dioxide levels in the atmosphere increased at about 0.6 0.1 ppm per year. However, the yearly growth rate between 2009 and 2018 was 2.3 ppm. [These figures are included in the American Meteorological Society's State of the Climate Report, which was published in late summer, along with the previous year's final global average.] The rate of rising in atmospheric CO2 during the last 60 years has been approximately 100 times faster than last natural increases, such as those observed at the end of the last ice age 11,000-17,000 years ago (Eskandari Damaneh et al., 2022).

4.7.1 What is the Relevance of Carbon Dioxide?

As a greenhouse gas, carbon dioxide absorbs and reflects heat. The Earth's land and ocean surfaces emit thermal infrared radiation (heat) after being warmed by sunlight (heat). Greenhouse gases, in contrast to oxygen and nitrogen, which make up the majority of our atmosphere, store heat and slowly release it, analogous to bricks in a fireplace after the fire has been out. Without this natural greenhouse effect, the average yearly temperature on Earth would be much lower than freezing, rather than approximately 60 degrees Fahrenheit. On the other hand, greenhouse gas emissions have thrown the Earth's energy balance off, trapping more heat and raising the global average temperature.

Carbon dioxide is the world's most abundant long-lived greenhouse gas. It absorbs less heat per molecule than methane and nitrous oxide, but it is more productive and lasts far longer in the atmosphere. Carbon dioxide emissions are responsible for more than two-thirds of the total energy imbalance causing global warming.

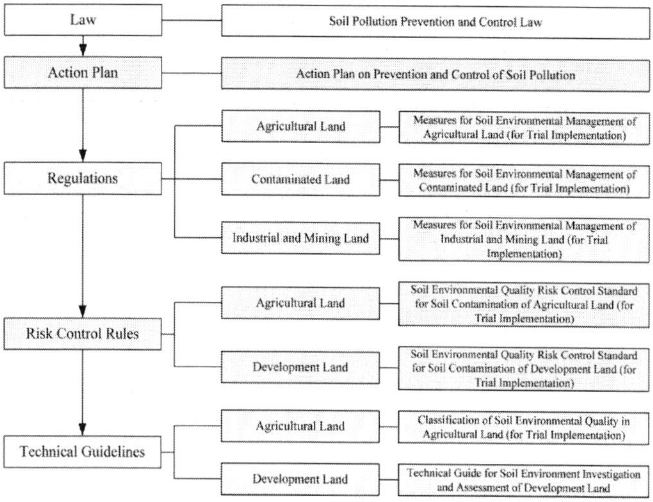

Figure 4.9: China's Fight against Climate Change and Environmental Degradation.

Source: Li T, Liu Y, Lin S, Liu Y, Xie Y. Soil Pollution Management in China: A Brief Introduction. Sustainability. 2019; 11(3):556. https://doi.org/10.3390/su11030556

Carbon dioxide has a huge influence on the Earth's system because it dissolves in the ocean like the fizz in a soda can. When it comes into touch with water molecules, it produces carbonic acid, which lowers the ocean's pH. (Acidification) The pH of the ocean's surface waters has fallen from 8.21 to 8.10 from the commencement of the Industrial Revolution. The phrase "ocean acidification" refers to the pH drop that has happened.

Although a 0.1-unit drop in pH may not appear to be significant, the pH scale is logarithmic, which means that a one-unit drop in pH results in a tenfold increase in acidity. There is a 30% rise in acidity for every percentage point increase in acidity. The ability of marine creatures to collect calcium from the water in order to form shells and skeletons is hampered when water acidity rises (Zhang & Huang, 2015).

Natural increases in carbon dioxide concentrations have produced periodic elevations in Earth's temperature during ice age cycles during the last million years or more. Warm periods (interglacial) start with a little increase in sunlight induced by a wobble in the Earth's rotation axis or orbit around the Sun. Warm times (interglacial) can extend hundreds of years.

The temperature climbed by merely a few degrees Celsius with a slight increase in sunlight. Carbon dioxide was produced as the water heated up, much like a soda can collapsing on a hot summer day. Global warming is accelerating faster than ever before due to rising carbon dioxide levels in the atmosphere.

Carbon dioxide levels have never exceeded 300 parts per million in the last million years, according to air bubbles trapped in mile-thick ice cores (and other pale climate evidence). According to the United Nations Environment Programme, pre-industrial carbon dioxide levels were roughly 280 parts per million on a worldwide scale (UNEP).

Global atmospheric carbon dioxide has already surpassed 315 parts per million during the first year of continuous monitoring at the Mauna Loa Volcanic Observatory, which began in 1958. The daily average carbon dioxide measurement at Mauna Loa surpassed 400 parts per million for the first time in the observatory's history on May 9, 2013. For the first time in less than two years, the worldwide level of ozone surpassed 400 parts per million in 2015, marking the highest level in more than a decade. If global energy consumption continues at its current rate, atmospheric CO2 levels are predicted to exceed 900 parts per million by the end of the century (Beilicci, Beilicci &Stefănescu, 2014).

4.8 WATER AND AIR POLLUTION

Soil contamination occurs when components, typically chemicals, are introduced into soils in an unnatural or too high quantity, posing a risk to humans or other living things. For instance, some feel pesticides are appropriate as long as their effect does not surpass the desired outcome, while others believe pesticides and chemical fertilizers should be avoided at all costs. It is difficult to identify soil contamination precisely because opinions on what constitutes a pollutant vary widely.

Both organic (carbon-based) and inorganic (non-carbon-based) contaminants are necessary for the efficient functioning of soils. Fuel hydrocarbons, polynuclear aromatic hydrocarbons (PAHs), polychlorinated biphenyls (PCBs), chlorinated aromatic compounds, detergents, and pesticides for organic pollutants are the most significant chemical categories (Shaner, 2017). Among the inorganic species (radioactive compounds) are nitrates, phosphates, heavy metals (including cadmium, chromium, and lead), as well as inorganic acids and radionuclides (radioactive substances). Numerous reasons contribute to the creation of these contaminants, including agricultural runoff, acidic residues, industrial waste products, and radioactive fallout.

Figure 4.10: Water Pollution and Soil Pollution.

Source: By Biswarup Ganguly - Own work, CC BY 3.0, https://commons.wikimedia.org/w/index.php?curid=18178073

If dangerous compounds flow into groundwater or polluted runoff reaches streams, lakes, or seas, soil contamination can lead to water pollution. Soil generates volatile compounds, which are discharged into the atmosphere and contribute to air pollution. Ammonia volatilization and denitrification are the two methods through which nitrogen escapes. When organic matter in the soil degrades, sulfur dioxide and other sulfur compounds are released, resulting in acid rain. The most crucial soil pollutants discovered in sewage are heavy metals and potentially hazardous chemicals. Heavy metals are present in sewage sludge, and if applied consistently or in large quantities, the treated soil may accumulate heavy metals, leaving it incapable of supporting plant life (Yakupoglu, Rodrigo-Comino &Cerdà, 2019).

Non-water soluble compounds also infect plants growing in contaminated soils, and they tend to accumulate higher up the food chain. The insecticide DDT was prohibited in the United States due to its propensity for getting into the soil on worms or fish and then on birds and their eggs. This occurred because higher-level species devoured animals polluted with the pesticide through their ingestion of plants and r microbes.

DDT concentrations in Lake Michigan are as follows:

Water contains two parts per trillion (ppt) (ppt).

There are fourteen parts per billion (ppb) (ppb).

Four hundred ten parts per billion (ppb) in amphipods (small water fleas and related animals) (tiny water fleas and similar creatures).

3–6 parts per million (ppm) in salmon and trout like coho and lake trout.

Up to 99 parts per million (ppm) in the top-of-the-food-chain herring gulls.

Over the last fifty years, one of the most severe concerns affecting scientists and the general public has been the environment's ever-increasing pollution. The constant release of synthetic organic compounds into natural ecosystems has been hastened by greater industrialization of agriculture, expansion of the chemical industry, and the need to offer inexpensive energy sources. As a result, the atmosphere, bodies of water, and many soil ecosystems have been poisoned with a vast array of hazardous substances (Blake et al., 2021). These compounds can be dangerous to humans and other animals in significant concentrations or extended exposure. In humans and other animals, these include acute toxicity, mutagenesis (genetic abnormalities), cancer, and teratogenesis (congenital impairments). Certain synthetic hazardous substances display resistance to physical, chemical, or biological degradation, constituting a severe environmental concern.

Polluted soils are decontaminated utilizing a range of in situ (on-site, inside the ground) and off-site (removal of contaminated soil for treatment) procedures. None of these methods is optimum for soil remediation, and a mix of them may be necessary to maximize the cleanup effort.

The most widely utilized way of decontaminating contaminated soils is to remove them and dispose of them in landfills or burning. However, these remedies frequently produce additional problems: landfilling holds toxic soil without purifying it, and incineration eliminates dangerous organic molecules from the ground but subsequently releases them into the air, resulting in air pollution.

Numerous soil washing strategies have been developed for the removal and recovery of heavy metals, including physical approaches such as attrition scrubbing and wet-screening and chemical processes such as treatments with organic and inorganic acids, bases, salts, and chelating agents (Rao,2014).

For example, hydrochloric, nitric, phosphoric, and citric acids are applied to remove radionuclides and dangerous metals. Other examples are:

- Carbonate of sodium.
- Hydrogen peroxide.
- EDTA and DTPA are chelating agents.

As with the preceding ones, the concern with these methods is that they create secondary waste products that may require additional hazardous waste treatment. In contrast to the prior techniques, in situ methods are immediately deployed at the polluted location. Because no soil is excavated in this case, the chance of inflicting environmental harm is negligible. In situ biodegradation is the technique of actively enhancing the population and activity of naturally occurring microorganisms. The bacteria then contribute to the breakdown of soil pollutants. Numerous environmental, chemical, and managerial variables influence soil pollutants' biodegradation, including moisture level, pH, and temperature, the microbial population present, and nutrient availability. Aerobic soil conditions and soil pH encourage biodegradation in the neutral range (between pH 5.5 and 8.0), with an ideal pH of around seven and a temperature of between 20 and 30°C. These physical features can be adjusted to boost the capacity of microorganisms to digest chemical pollutants. Bioremediation looks to be the least destructive and ecologically acceptable form of decontamination (Webb, Herrick, &Duniway, 2014).

4.9 POPULATION DISPLACEMENT

"Developed countries cannot remain immune to suffering and disaster in developing countries: large numbers of environmental refugees have already entered OSCE countries, the majority of them illegally, and today's flow will undoubtedly be considered a trickle in comparison to the floods that will erupt in the coming decades," according to the statement. Among those recognized are the University of Oxford, Green College, and Professor Norman Myers. "This is a tough issue to handle because global institutions are already overburdened with obligations related to historically recognized refugees, as defined by the 1951 Refugee Convention." "This new species of "refugee" must be acknowledged, accepted, and assimilated into society as soon as possible within the framework of international law." According to Dr Hans van Ginkel of the United Nations University, it is reasonable that as the world contends with the effects of climate change, the number of people fleeing unlivable environmental conditions would expand enormously. This new "refugee" category must be acknowledged under international treaties.

We must improve our capacity to fulfill the needs of individuals fleeing other unsustainable environments." According to Dr. Janos Bogardi of the United Nations University's Institute for Environment and Human Security, sustainable development has the potential to lessen susceptibility to climate change through improving adaptive capacity and resilience. "By building adaptive capability and resilience, sustainable development has the potential to lessen susceptibility to climate change" (UNU-EHS). Only a few existing sustainability programs, however, explicitly address the effects of climate change or the development of capability. The IPCC (Intergovernmental Panel on Climate Change) is a non-governmental organization that undertakes research on global warming and climate change." The Intergovernmental Panel on Climate Change (IPCC) said some years ago that "the most severe ramifications of climate change may be those on human migration, when millions of people are displaced as a result of coastal erosion, floods, and drought." Climate change, according to a recent study, will have a severe influence on both natural and social systems in the near future. Pessimists believe that a big number of people will be compelled to move, generating political upheaval, bloodshed, massive human misery, and imposing a significant load on recipient civilizations. A number of contentious words have evolved as a result of discussions about the link between environmental deterioration and forced migration (Van Leeuwen et al., 2019).

To mention a few, they include climate refugees, climate change refugees, environmentally displaced persons (EDP), disaster refugees, environmental displacement, eco-refugees, ecological displaced persons, and ecologically refugee-to-be (ERTB). These notions have no substance in international refugee law, as environmental considerations do not constitute a legal reason for international protection. They are not legal positions imposing duties on governments, but rather descriptive statements describing the reality. Most of the time, fundamental assessments and established definitional labels are employed to steer discussions on the validity of claims. Because there is no precise definition for commonly used terms, there are fears surrounding the emotionally charged issue of migration, estimates of the scale of climate-induced displacement vary widely, and there is a lack of dialogue between ecologists and social scientists, drawing connections between environmental change and forced migration is difficult and contentious. Some opponents believe that research on the link between climate change, extreme environmental events (EEEs), and migration patterns is still in its early phases.

Figure 4.11: People's displacement: Villagers fleeing gunfire in a camp for internally displaced persons during the 2008 Nord-Kivu war.

Source: By Julien Harneis - https://www.flickr.com/photos/julien_harneis/3009852745/sizes/l/, CC BY-SA 2.0, https://commons.wikimedia.org/w/index.php?curid=5236657

However, there is a rising recognition of the significance of adopting methods to prevent environmental-induced migration and conflict. The extent and complexity of the environmental migration problem are difficult to predict, especially considering the variety of causes of forced migration. Therefore, it is critical to investigate the extent to which environmental degradation is a root cause of migration and conflict, to address the issue of environmentally-induced migration as soon as possible, and to develop consistent policies that account for vulnerability to environmental, economic, and political changes, all of which must be supported by rigorous scientific and academic research. The issues of humanitarian protection must be addressed within a long-term development policy framework that stresses resilience, livelihoods, and adaptability (Tsoraeva et al., 2021). This viewpoint, rather than more ominous assumptions about the likelihood of forced displacement and violence, is an important part of this strategy research. Coercive environmental remediation is, of course, not a new issue. Land scarcity and environmental deterioration have caused waves of migration and conflict throughout history. Throughout human history, migration and population movement have been key adaptive acts. As a result, it has always been difficult to discern between "environmental refugees" and "economic migrants." The urge to evacuate a disaster-stricken area, combined with the economic allure of a more promising place, may result in a migratory decision on rare occasions.

Following the 1927 Mississippi Delta flood, three million people fled the Dust Bowl in the 1930s, including 700,000 primarily poor black people. Their initiatives reflected a confluence of demands and intentions in some cases. However, both the present and the near future have two characteristics. To begin, global environmental change and its possible consequences, such as forced migration, are current occurrences with long-term consequences. Second, environmental change is inextricably linked to human adaptability. We accept these realities and urge for policy solutions to be implemented at all levels of government. Its basic idea is that proactive activities that promote resilience, flexibility, and long-term livelihood sustainability are the most effective way to solve a wide range of environmental issues (Tully et al., 2015). This policy research investigates the origins of the phrase "environmental refugee" before digging into the accuracy of environmental relocation projections. The researchers then devised unique approaches for evaluating the potential breadth of migration and, as a result, the scope for policy intervention. The authors offer data from nations and areas with the most susceptible populations to climate change, as well as proactive resilience, sustainability, and adaptation strategies for preventing environmental-induced migration. The link between environmental decline and conflict is next investigated. We shoulddd analyze potential legal and institutional reforms, calling for multifaceted methods to acknowledging and mitigating the human consequences of climate change. Policy suggestions are provided as a result of such research.

4.10 SOIL CONTAMINATION AND SPECIES EXTINCTION

Biodiversity refers to the diversity of plant and animal life on the globe. The terrestrial environment's tropical and subtropical zones have the highest biodiversity on the planet. It's no surprise that the Amazon jungle is the world's most biodiverse place, as high biodiversity is linked to increased rainfall. Despite this, many unique ecosystems, especially in dry environments, may be capable of maintaining a high level of biodiversity. With over 5000 plant species, roughly 40% of which are unique, the Succulent Karoo in South Africa may be the dry region with the highest desert biodiversity.

In and of itself, biodiversity loss might be considered a kind of land degradation.

For example, overhunting has resulted in the extinction of the majority of essential herbivores, carnivores, and primates in tropical forests, creating an "empty forest." This might set off a chain reaction that affects all trophic levels. The reintroduction of wolves into Yellowstone National Park in the United States is an example of how the extinction of a large predator may have far-reaching consequences. Elk numbers increased in the absence of wolves, trampling flora and creating erosion along riverbeds (Blake et al., 2018). Human activities have resulted in the targeted extinction of individual species in rare cases. In South Africa, for example, medical plant extraction has put 82 plant species on the verge of extinction. Biodiversity strengthens ecosystems, whereas loss of biodiversity jeopardizes ecosystem processes and services. Ecosystems' capacity to adapt to climate change may be particularly crucial.

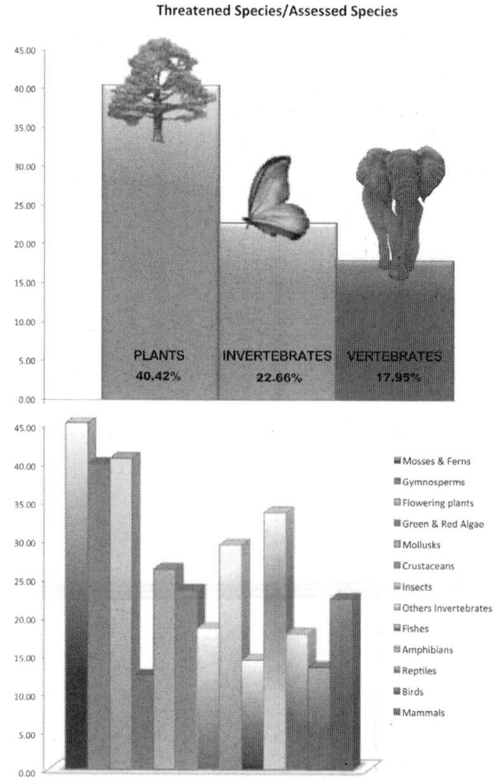

Figure 4.12: Sixth mass extinction of wildlife accelerating.

Source: Palombo MR. Thinking about the Biodiversity Loss in This Changing World. Geosciences. 2021; 11(9):370. https://doi.org/10.3390/geosciences11090370

Biodiversity loss is exacerbated by land degradation. As ecosystems deteriorate, they lose their ability to retain biodiversity, and biodiversity in highly degraded environments is typically diminished or moved to early successional species. Overgrazing kills perennial grasses, allowing annual grasses and weedy forbs to take their place, especially exotic invading species. Soil erosion occurs, as a result, making cow grazing less appealing. The most significant influence on biodiversity is the conversion of natural or semi-natural vegetation to agricultural fields or human settlements, which frequently results in a near-total loss of indigenous species.

The Millennium Ecological Assessment promoted ecosystem services and their importance to humankind. Even though the relationship between biodiversity and ecosystem services is complicated, there is widespread consensus that biodiversity is necessary for long-term and sustainable ecosystem service fluxes. With increased aridity predicted due to climate change, dry land species' genetic diversity may be crucial for breeding new types that resist higher temperatures and lower rainfall (Telles, Guimarães, &Dechen, 2011). The value of biodiversity in arid environments is sometimes ignored since dry places are less productive than more wet areas. Regardless, these dry regions house a disproportionately large share of the world's population and offer significant food and other ecological services.

4.11 BIODIVERSITY CONSERVATION AND POVERTY

"Humans are forced to slaughter the goose that lays the golden eggs by poverty."

In most, if not all, dry land ecosystems, poverty and biodiversity overexploitation have a destabilizing relationship. Overexploitation of woody plants is a classic example when poverty forcing people to utilize inexpensive biomass-based fuels for cooking, heating, and other types of energy delivery rather than superior technologies. As a result, cities and villages are bordered by vast swathes of devastation, with trees, shrubs, and dwarf plants wiped off in a matter of a dozen kilometers. Charcoal is replacing fuel wood since it is easier to transport and has an impact that may be felt for several kilometers. As a result, many of the previous woody vegetation that offered ecosystem services, including food, medicine, construction materials, textiles, crafts, and fertilizer, have been depleted.

The planet is losing species at an unprecedented rate right now. Human-caused extinction accounts for the majority of such species extinction. Land conversion, such as from natural woods to agricultural crop fields, has posed a serious threat to biodiversity in recent decades. Despite the global objective of preserving at least 17% of each habitat in protected areas, several plant species have very low levels of official protection. The natural habitat has been destroyed in some cases. Species are lost due to the 'island biogeographic' impact when natural habitats are disrupted (Salesa&Cerdà, 2020). The fragmented habitats obstruct species migration, and the remaining areas may become too small to sustain healthy populations of some species. Land degradation in natural environments has the potential to reduce biodiversity severely. Many cow ranching systems, for example, have eliminated the majority of big indigenous herbivores, and overgrazing may also significantly alter the plant species mix in the herbaceous layer. Invasive alien species negatively influence native biodiversity, displacing native species and changing the environment's character. Pollution has a negative influence on biodiversity, particularly nitrogen deposition.

The effect of climate change, which is expected to result in unprecedented species extinction rates in the future decades, is heaped on top of that. Species and ecosystems will be forced to shift to warmer climates due to climate change. As the globe warms, people migrate further from the equator or higher up mountains in search of cooler climes. Land deterioration and change will make migration for this species much more difficult, hastening its extinction.

4.12 ECONOMIC IMPACT

We are losing more than just soil as a result of erosion. According to a new study, soil erosion reduces agricultural productivity and increases water usage, costing the global economy $8 billion.

Soil erosion is becoming more severe as a result of climate change. Certain countries are suffering as a result. There are many more dry days intermingled with significant rains due to climate change. During big storms, this creates conditions where dirt is quickly swept away. By removing vegetation that helps hold the soil together, human land-use change exacerbates water-induced soil erosion. Steep terrains, such as hills, may worsen the problem.

The RUSLE and MAGNET models were used by researchers from the European Union Joint Research Centre to analyze economic losses caused by soil erosion. They determined that the global economy loses $8 billion every year. Global food production has decreased by 33.7 million tons. Researchers were able to estimate losses from structural economic change due to changes in primary resources, such as land quality, in this study, which is a first (Rickson, 2014). Crop less was previously just estimated and multiplied by market prices. This was also the first time this technology has been used worldwide.

Countries with high rates of erosion and extensive areas of agricultural land are particularly vulnerable. On more than 70% of their arable land, most Caribbean countries, Brazil, Central African countries, and sections of Southeast Asia are facing severe erosion. In comparison, severe decay affects barely 3% of arable land in Australia, Canada, Saharan nations, Russia, and the majority of the European Union. To put this in context, 24 percent of the world's arable land is rapidly degrading.

These findings indicate that the least responsible countries for climate change will bear the brunt of the consequences. Increasing sea levels, disease outbreaks, and, most lately, the loss of arable land have all proved this.

India and Indonesia are "the largest losers in percentage terms related to soil erosion" compared to regions with a more excellent agricultural base, with losses of up to 0.1 percent of GDP.

Kenya is primarily a farming country, with over 12 million people living on degraded soil. Unfortunately, the yield of the country's food crops has not risen in lockstep with population expansion. Soil erosion, low productivity, loss of agro-biodiversity, and soil nutrient depletion limit Kenya's agricultural output development. Land exploitation would result in significant land degradation without adequate compensation in soil and water conservation. Rural livelihoods will be destroyed, water supplies will be decreased, and animal habitats will be affected as a result (Chuchkalov, Fadeev & Alekseev, 2020).

Furthermore, based on milk and meat output losses and animal live weight losses, the yearly costs of rangeland degradation are estimated to be around $80 million. Furthermore, the annual expenses of "soil nutrient mining," which results in decreased yields for three Kenyan crops, namely wheat, maize, and rice, are projected to be around 270 million USD. The cost of taking action to repair LUCC-damaged lands is demonstrated to be four times cheaper than the cost of inaction over 30 years, implying that every dollar invested in land rehabilitation will return four dollars. This might provide compelling evidence for the urgent need to address land degradation. SLM investments are essential to reduce land degradation. According to our econometric findings, increasing farm families' access to SLM information and markets (input, output, and financial) will likely inspire them to participate in SLM.

The economics of land degradation has received little attention until recently. Even though many researchers have attempted to measure ecosystem services from a global to a micro level, investigations that look at the total costs of land degradation are still uncommon. This section aims to assess the value of various land resource modeling and valuation methodologies utilized in previous studies, as well as the types of data used in these assessments for multiple types of land resource and management appraisals. We also look at the advantages and disadvantages of various valuation approaches used in soil erosion research and how they are used. We feel that new models are needed to improve the study's accuracy in calculating the economic consequences of land degradation while taking the regional variety of biophysical and economic factors into account.

In 2008, it was estimated that deforestation and land degradation would cost the world economy EUR 1.5–3.4 trillion per year, or 3.3–7.5 percent of global GDP. This includes astonishing yearly grain losses of USD 1.2 billion. Annual soil loss of 75 billion tons from agricultural land owing to degradation costs the globe more than USD 400 billion, with the United States alone losing USD 44 billion due to soil erosion. Using SLM approaches, reaching 95 percent of ideal agricultural yields may result in an additional 2.3 billion tons of food output every year, worth USD 1.4 trillion (Cerda et al., 2018).

In 2000, nearly 2 billion people lived in arid areas, accounting for 41 percent of the earth's terrestrial surface, with 90 percent being in developing nations. Water scarcity may result in the displacement of 700 million people by 2030, with up to 50 million moved in the next decade attributable to desertification alone. In degraded desert ecosystems, human well-being and poverty rates are high, a situation compounded by quick population expansion. Dry lands, for example, increased at the greatest speed of any ecosystem assessed in the Millennium Ecosystem Assessment throughout the 1990s, with an average rate of 18.5 percent. Other advantages to the environment are diminishing. Land degradation exacerbates deforestation, which yearly burns 13 million hectares of forest. Changes in land cover throughout the previous two decades have diminished the yearly flow of ecosystem services by USD 4–20 trillion. Land degradation is predicted to result in a loss of ecosystem services of between USD 6.3 and 10.6 trillion per year. Ecosystem services are estimated to be valued between 10 percent and 17 percent of world GDP (USD 63 billion in 2010). Changing global regulations to allow for sustainable land management might result in a $ 75.6 trillion yearly benefit. Small-scale valley bottom irrigation has already generated returns of more than 40%.

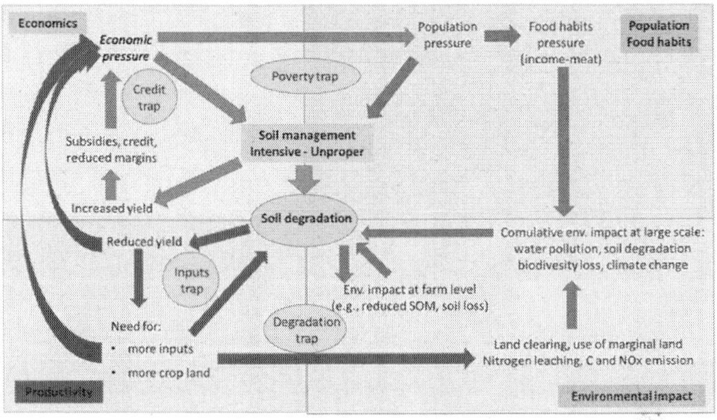

Figure 4.13: Socio-economic impacts of soil pollution.

Source: Gomiero T. Soil Degradation, Land Scarcity and Food Security: Reviewing a Complex Challenge. Sustainability. 2016; 8(3):281. https://doi.org/10.3390/su8030281

CHAPTER 5

HOW CONTAMINANTS AFFECT SOIL FUNCTIONS

CONTENTS

5.1 Interaction With the Environment 115
5.2 Mapping Soil Functions 117
5.3 Categories of Contaminated Lands 119
5.4 Microplastics 121
5.5 Hydrocarbons 124
5.6 Electronic Waste 128
5.7 Sites of Special Concern 131
5.8 Exposure to Contaminant 133
5.9 Pesticides and Herbicides 134
5.10 Conclusion 137

The introduction of xenobiotic chemicals and perhaps other changes in the organic soil ecosystem cause soil contamination, soil pollution, or land pollution as a component of land degradation. It is commonly triggered by industrial production, agrochemicals, or inappropriate waste disposal. Petroleum hydrocarbons, PAHs (such as naphthalene and benzopyrene, solvents, insecticides, lead, as well as other heavy metals, are by far the most frequently involved compounds. Contamination is proportional to the degree of industrialization and the concentration of chemical substances. Concerns about soil pollution originate mostly from health concerns associated with direct contact with contaminated soil, vapor from toxins, or secondary contamination of water sources inside and beneath the soil (Cogo, 2022).

Mapping contaminated soil locations and the subsequent cleanups are time-consuming and costly activities that necessitate skills in geology, hydrology, chemistry, computational modeling, as well as an understanding of industrial chemistry's history.

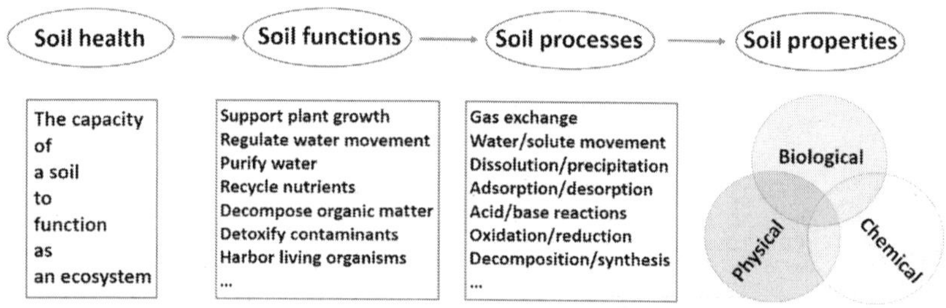

Figure 5.1: Fundamental soil functions in ecosystems.

Source: Guo M. Soil Health Assessment and Management: Recent Development in Science and Practices. Soil Systems. 2021; 5(4):61. https://doi.org/10.3390/soilsystems5040061

The magnitude of contaminated land is well known in North America and Western Europe, with many governments in these regions having a legislative framework in place to detect and address this environmental concern. Notwithstanding the fact that some emerging economies have developed tremendous industrialization, they continue to be less closely regulated.

Soil is a limited resource, which means that its depletion and degradation cannot be recovered inside a human lifetime. Soils have an impact on the foods we consume, the water we use, the air we breathe, our health, and the health of all lifeforms. We will not be able to cultivate food lacking healthy soils. Indeed, it is believed that nearly 100% of our food is generated explicitly or implicitly on our soils.

Soils that are healthy are the cornerstone to food security and a viable tomorrow. They aid in the sustainability of food production, the mitigation and adaptation to changing climate, the filtering of water, the tolerance to floods and droughts, and far more. However, an unseen threat is jeopardizing soils and everything they have to offer. Pollution of the soil sets off a chain reaction. It affects soil biodiversity, decreases soil organic materials, and diminishes soil's potential to behave as a filter. Also, it pollutes the water held in the soil and groundwater and generates a nutritional disequilibrium in the soil. Heavy metals, lasting organic contaminants, and new contaminants – such as pharmaceuticals and hygiene products items – are among the top prevalent soil pollutants (Poesen, 2018). Soil is a sophisticated collection of minerals, biological matter, water, and numerous organisms, as you've discovered from other pages. Soil was originally an uncontaminated substance that covered the earth. However, in certain regions, humans have purposefully or unintentionally thrown toxic materials onto it. The waste has the potential to harm the land as well as human, plant, and animal welfare. This section discusses the most prevalent contaminants, how they became present, and how they can endanger human health.

A soil contaminant is defined as any material in the soil that surpasses naturally existing levels and constitutes a risk to human health. Arsenic, for instance, exists naturally in certain soils. However, if an individual sprays some pesticides on their lawn, it may create soil contamination. Lead is also extremely hazardous, but it exists naturally in certain soils. It was utilized in gasoline till the 90s and is still present in soil now.

Urban regions and previous industrial sites pose the greatest danger of soil pollution. If you are concerned about the health of the soil system of your residence or property, it is important to have a soil test performed. In fact, most soil is absolutely safe for playing, cultivation, and leisure, but it's always better to be cautious than sorry (Eskandari Damaneh et al., 2022).

Pesticides, hydrocarbons, radon, asbestos, arsenic, and creosote are all main sources of pollution in urban soils. Human activities are mostly responsible for soil contamination in metropolitan areas. Industry, commercial dumping, urban expansion, local garbage disposal, and inappropriate pesticide or chemical use are a few illustrations.

High car and truck traffic, and even just a single automobile, can pollute soil: A gleaming puddle under where you park your vehicle is just one example. That is a petroleum-based product that when precipitation occurs, will seep into the soil. When soil becomes contaminated with any of these pollutants, the natural ecology suffers. Many of these compounds are poisonous to plants in the same way that they are to people. Furthermore, because soil is the "planet's kidney," toxins can pass through it and into our water supply. It's easy to see why soil pollution really is a contentious issue!

The location and amount of pollution introduced to soils will have a big impact on how the contamination travels across an area. The soil types will indeed influence its spread. Certain toxins, for instance, could enter groundwater aquifers more quickly in sand than in clay. This is due to coarse-grained sandy soil types' quicker penetration rates. Pollutants can be held tightly in fine-grained clay soils or organic material in surface soils, which means they will collect if left unchecked (Fenta et al., 2021).

Some human activity may occur concurrently. Some new home developments, for example, are developed on former industrial grounds. Those soils may include subsurface soil contaminants, which are made visible when structures – or roads that carry them – are built. Contaminants that have recently reappeared may then be dispersed all across the urban ecology via wind (as dust) and runoff water. Contaminants might be preserved at a location, posing an on-site danger of contact if foodstuffs are produced or kids play in the ground.

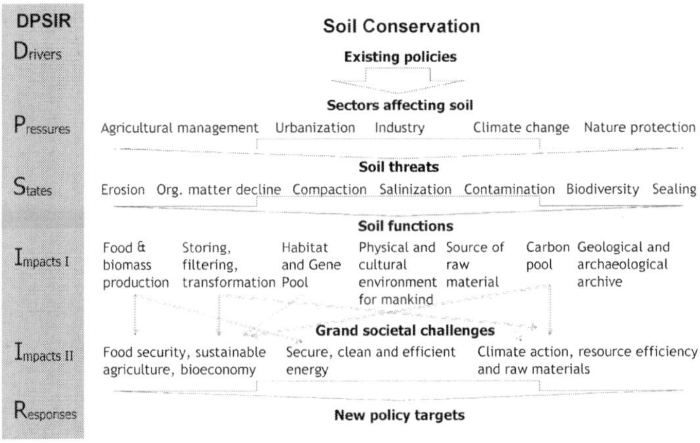

Figure 5.2: Correspondence between soil functions and ecosystem services.

Source: Glæsner N, Helming K, De Vries W. Do Current European Policies Prevent Soil Threats and Support Soil Functions? Sustainability. 2014; 6(12):9538-9563. https://doi.org/10.3390/su6129538

Soil pollution is harmful to the ecosystem and has ramifications for all living things that come into contact with it. Unsustainable agriculture methods that deplete soil organic matter might allow contaminants to enter the food chain. Contaminated soil, for instance, can discharge toxins into groundwater, that accumulates in plant cells and is transferred on to grazing species, birds, and, lastly, people who eat the flora and fauna (Pijl et al., 2020). Contaminants in the soil, groundwater, and food system can create a wide range of illnesses and increased mortality in people, ranging from short-term acute impacts like intoxication or diarrhea to long-term chronic effects including cancer. Soil pollution has enormous economic implications in addition to its environmental impact since it reduces crop production and quality. Preventing soil pollution ought to be a primary priority on a global scale. Because the heavy bulk of pollutants is released by human activity, we are immediately accountable for implementing the required reforms to guarantee a less contaminated, more secure tomorrow.

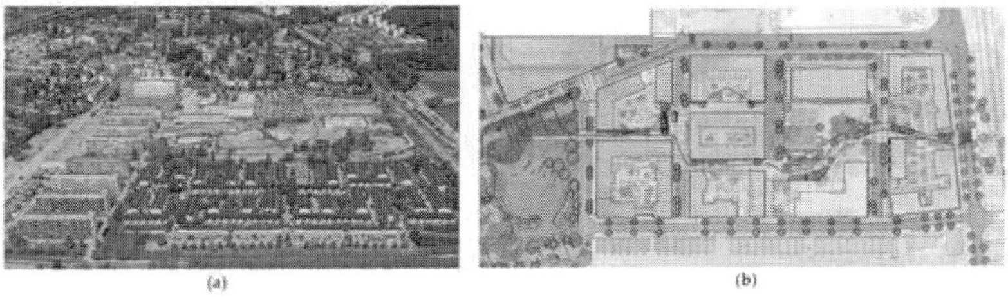

Figure 5.3: Urban Development.

Source: Brokking P, Mörtberg U, Balfors B. Municipal Practices for Integrated Planning of Nature-Based Solutions in Urban Development in the Stockholm Region. Sustainability. 2021; 13(18):10389. https://doi.org/10.3390/su131810389

Soils must be acknowledged and respected for their productive capacities, and also their role in food and nutrition security and the preservation of critical ecosystem services. Here are a few examples of why soil pollution should not be underestimated:

- Soil pollution has an impact on everything. The food we consume, the water we sip, and the air we breathe are all reliant on healthy soil for our health and health of all life forms on earth. The nutritional content of a plant's tissues is directly proportional to the soil's nutrient content and the process of exchanging nutrients and water with the plant's roots.

- Soil contamination is imperceptible. Today, one-third of global soils are somewhat or severely deteriorated as a result of erosion, soil organic carbon loss, salinization, compaction, acidity, and chemical contamination. It takes around 1000 years to build one centimeter of topsoil, implying that we will not be capable of creating more soil throughout our lifetime. All we can see is what we see. Soil contamination, on the other hand, is putting additional strain on soils. The present speed of soil degradation jeopardizes future generations' abilities to fulfill their most basic requirements (Pham et al., 2018).
- Soil contamination reduces soil's ability to filter. Contaminants are filtered and buffered by soils. The ability of soil to withstand the strain of pollution is limited. Contaminants leak into other sections of the environment, including our food chain, if the ability of soil to shield us is surpassed.
- Soil pollution has an impact on food security by lowering crop production and quality. Food that is safe, nutritional, and of high quality can only be prepared if our soils are in good health. It won't be possible to attain #ZeroHunger until we have healthy soils.
- Poor farming practices can lead to soil contamination. Unsustainable farming practices deplete soil organic matter, jeopardizing the soil's ability to breakdown organic contaminants. This raises the possibility of contaminants being discharged into the ecosystem. Intensive agricultural production has exhausted soils in several states, endangering our tendency to preserve production in these places in the future. As a result, sustainable agricultural processing methods have become critical for reforming the systems that encourage soil degradation and maintaining present and future international food security.
- Soil pollution can endanger our health. Antibiotics, which are commonly employed in agricultural and human healthcare, are extensively discharged into the ecosystem after being expelled from the animal to that they were provided. These medicines have the potential to permeate our soils and proliferate across the environment. This results in antimicrobial-resistant bacteria, reducing the effectiveness of antibiotics. Antimicrobial-resistant microorganisms cause around 700,000 fatalities per year. If not

addressed, it will cause more deaths than cancer by 2050, costing more than the existing international economy (Pietro & Mahajan, 2022).

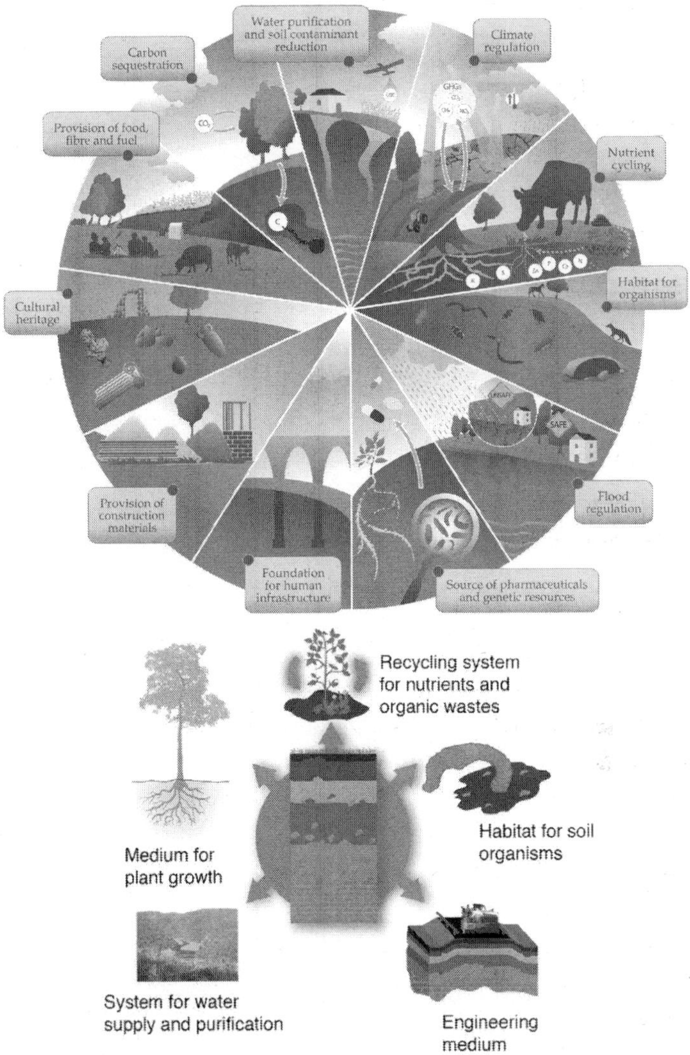

Figure 5.4: Functions of the Soil.

Source: Baveye PC, Schnee LS, Boivin P, Laba M and Radulovich R (2020) Soil Organic Matter Research and Climate Change: Merely Re-storing Carbon Versus Restoring Soil Functions. Front. Environ. Sci. 8:579904. doi: 10.3389/fenvs.2020.579904

With a predicted worldwide population of 9 billion by 2050, our present and future food security is dependent on our capacity to enhance harvests and quality of food. Soil contamination has a severe influence on all of us and is being recognized as one of the major risks to soil functions globally.

We must understand the origins of soil pollution in order to develop and execute solutions. It is up to us to safeguard and conserve the soil. Adopting sustainable food decisions, appropriately recycling hazardous goods such as batteries, recycling and composting at households to limit the quantity of trash that reaches landfills, and more appropriately handling antibiotic waste are just a few instances of how we can contribute to the answer. We need to encourage sustainable agriculture techniques in our communities on a greater scale (Frankl et al., 2016).

Healthy soil is a valuable, non-renewable resource that is progressively under threat from damaging human actions. We are accountable for the soils that give us food, water, and air, and we must act now to preserve healthy soil for a viable and food-secure tomorrow.

Soil functions are basic soil characteristics that are significant for a variety of agronomic, ecological, nature preservation, landscape design, and urban activities. Soil can serve several purposes, including those connected to ecological systems, crop production, environmental condition, raw material supply, and architectural foundation. The following are the six most important soil functions:

- Production of food and other biomass
- Interaction with Nature
- Biological environment and gene pool
- Raw material source
- Natural and cultural heritage
- A venue for anthropogenic structures

Plant roots are held in place by soil. It creates a reliable haven for plants to dwell in while also storing and giving nutrients to them. Soil also operates by permitting carbon dioxide to exit and clean oxygen to reach the root zone, thereby preserving the volume and purity of air. Soil pores can also collect water and keep it for when plant roots require it (Priori et al., 2020). The soil also significantly influences temperature fluctuations, ensuring that the roots can operate properly. Fertile soil will also include absorbed mineral nutrients that will aid in plant growth. The confluence of these processes promotes plant growth for food and biomass generation.

5.1 INTERACTION WITH THE ENVIRONMENT

The soil is influenced environmental exchanges like managing water sources, utilization, pollution, and purification. They can screen, cushion, and alter elements that pass through the atmosphere, vegetative cover, and groundwater level. Soil communicates with its surroundings to change and breakdown waste substances into new ones. Soil works as a filter, capturing impurities through soil particles, and this is accomplished by filtering. Contaminants are collected by soil particles, and water in aquifers and rivers is cleaner as a result. Finally, it can collect enormous volumes of carbon as soil organic materials, lowering total carbon dioxide concentrations and so mitigating global climate change. Interaction with the environment

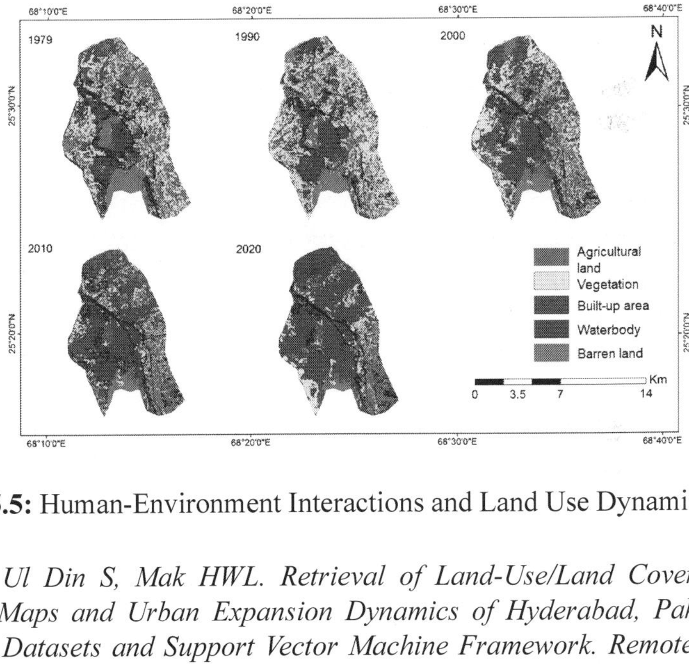

Figure 5.5: Human-Environment Interactions and Land Use Dynamics.

Source: Ul Din S, Mak HWL. Retrieval of Land-Use/Land Cover Change LUCC) Maps and Urban Expansion Dynamics of Hyderabad, Pakistan via Landsat Datasets and Support Vector Machine Framework. Remote Sensing. 2021; 13(16):3337. https://doi.org/10.3390/rs13163337

5.1.1 Biological Habitat and Gene Pool

Soils also function as a biological home and a genetic bank for a wide range of creatures. Soils are the ecosystems in which seeds develop; they supply heat, nutrients, and water, which can be used to nourish plants and animals. Many living species can benefit from soil's aid in the breakdown of deceased plants, animals, and organisms by converting their remnants into simple mineral compounds (Pimentel & Burgess, 2013).

5.1.2 Raw Material Source

Soil offers raw ingredients for human usage and has a direct effect on human health. Human food's content mirrors the character of the land in which it has been grown. An instance of dirt being used as a source of raw materials can be traced back to ancient ceramic manufacture. The Maya pottery exhibited characteristics acquired from the raw materials utilized, which were soils and sediments. Understanding the soil development process can assist define specific types of soil and reflect the mineral makeup of soil. Nevertheless, the natural region of fertile soil is finite, and because of rising farming, forestry, and urbanization pressures, harvesting soil as a raw resource must be regulated.

5.1.3 Physical and Cultural Heritage

Soil also has broader cultural purposes in that it is a component of the cultural zeitgeist of our brains as well as the literal environment around us. A strong affinity to home grounds or a special identity is a cultural trait that develops more strongly in some people. Soils have been around since the beginning of time, and they can play a role in defining how humans have traveled in the past. Soil also serves as an earth cover, protecting and preserving tangible relics from the past, allowing us to better comprehend cultural legacy. Furthermore, because soil is a key resource for human production, it has been a crucial indicator of where people reside.

Figure 5.6: Palmyra.

Source: By Bernard Gagnon - Own work, CC BY-SA 3.0, https://commons.wikimedia.org/w/index.php?curid=12153761

As a platform for man-made constructions: soil can be utilized as a raw material deposit and is commonly employed in building supplies. Roughly half of the world's population lives in dwellings made of soil. Soil properties must be hard and stable in order to offer a good foundation for the construction of roads and motorways. Furthermore, because these constructions are supported by soil, characteristics including its ultimate load, compressibility, durability, and ultimate strength must all be assessed. Evaluating the physical properties of soil allows for a more accurate deployment of engineering applications of soil (Nearing, Lane, & Lopes, 2017).

5.2 MAPPING SOIL FUNCTIONS

Mapping of soil characteristics is an augmentation of soil survey that uses maps of soil attributes in conjunction with auxiliary information (such as pedotransfer ratios and soil prediction models) to illustrate judgments about the unique functionality of soil mapping sections. Other roles of soil in environments include:

- Source of building materials (clay, sand, rocks)
- Carbon recycler
- Fiber production

Figure 5.7: An example of soil mapping.

Source: Nawar S, Buddenbaum H, Hill J. Digital Mapping of Soil Properties Using Multivariate Statistical Analysis and ASTER Data in an Arid Region. Remote Sensing. 2015; 7(2):1181-1205. https://doi.org/10.3390/rs70201181

5.2.1 Types and Extent

Manufacturing, mineral extraction, mine abandonment, national military operations, waste management, accidental spillage, illegal discharging, storage leaks, hurricanes, floods, pesticide utilize, and fertilizer application of fertilizers are all examples of intentional, unintentional, or naturally occurring processes and events that can cause land contamination (Nigussie et al., 2015).

Thousands of polluted sites of different size and significance exist around the country, ranging from abandoned structures in inner cities to enormous areas poisoned with toxic compounds from previous industrial or mining operations.

Contaminated lands include the following:

- Contaminated sites as a result of incorrect management or disposing of poisonous and hazardous chemicals and wastes.
- Locations where harmful chemicals may have been dumped as a consequence of biological catastrophes or terrorist acts
- Locations where incorrect management or disasters have caused in the discharge of toxic and harmful materials which are not wastes.

Many areas are monitored at the state level, especially the biggest and most substantially contaminated, but many more are only monitored at the state or municipal level. There is no single collection of thorough data that tracks the whole amount of contaminated land in the US.

In 2008, the EPA broadened the extent of its national surveillance efforts to encompass all sorts of sites under its jurisdiction, and also projections of the area ascribed to such sites. The number and state of polluted areas change often when new contaminated sites are located, recorded, and cleared away (for example, as a result of spills or natural catastrophes such as hurricanes).

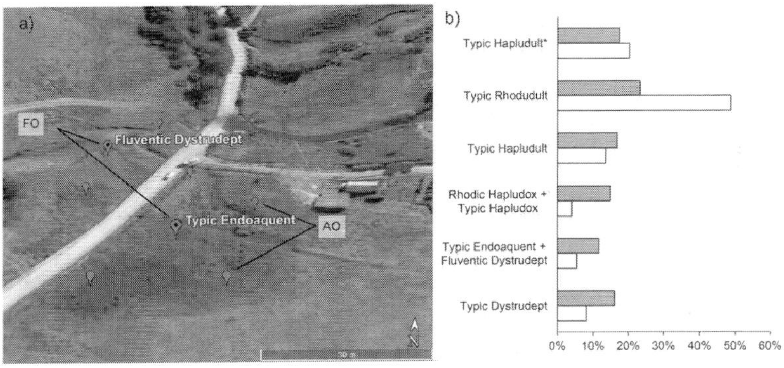

Figure 5.8: (a) Allocation of additional observations (AO) to field observations (FO) in Google Earth for the soil mapping unit (SMU) TypicEndoaquent+FluventicDystrudept.(b) Original and final proportions of the training datasets by SMU.

Source: Silva, B.P.C., Silva, M.L.N., Avalos, F.A.P. et al. Digital soil mapping including additional point sampling in Posses ecosystem services pilot watershed, southeastern Brazil. Sci Rep 9, 13763 (2019). https://doi.org/10.1038/s41598-019-50376-w

5.3 CATEGORIES OF CONTAMINATED LANDS

- National Superfund Priorities Industrial facilities, waste collection sites, mining and sedimentation sites, and federal infrastructure like mining sites, nuclear, biological, chemical, and standard arms production plants, and army facility industrial sites are all on the list of highly contaminated sites (e.g., those used for aircraft and naval ship maintenance).
- Resource Conservation and Recovery Act (RCRA) cleanup sites: These establishments are subject to RCRA cleanup owing to previous or current hazardous waste processing, storage, or disposal and have historical contaminated releases (Molla & Sisheber, 2017).
- Underground storage tanks/leaking underground storage tanks: Businesses, industries, gas stations, and numerous institutions store petroleum and dangerous materials in huge underground storage tanks, which can fail caused by faulty materials, setup, operating processes, or maintenance systems, resulting in soil contamination (Ettler, 2016).

- Unintended spill sites: Countless oil, gas, and chemical spills happen on land and in water every year as a result of a range of mishaps, such as transportation (e.g., railways, barges, trucks, tunnels) and facility discharges.
- Contaminated sites from natural disasters or terrorist activities: Disasters of any kind, whether natural or man-made, have the propensity to contaminate soils and create difficulties at already contaminated locations.
- Land contaminated with radioactive and other dangerous chemicals: As a consequence of operations linked with nuclear weapons development, testing, and research, many sites spanning a broad amount of land in the US are contaminated with radioactive as well as other harmful materials.
- Brownfields: Brownfields are properties where the existence or possible presence of a hazardous material, pollutant, or contaminant complicates development, rehabilitation, or reuse. Cleaning up and investing the money in these sites helps to maintain the environment, lowers blight, and relieves development demands on green areas and productive lands.
- Military bases and defense sites: The Department of Defense uses millions of acres of land that has been contaminated by hazardous material and pollutant discharges, spent munitions, armaments constituents, and explosive materials, and building destruction debris. Likewise, the Department of Energy has a number of sites that have been affected by discharges of contaminants and/or radioactive materials (Moges & Holden, 2007).
- PCB-contaminated sites: Polychlorinated biphenyls (PCBs) were extensively employed in very many commercial industries before the Toxic Substances Control Act, and substantial PCB contamination occurred from spills and discharges, as well as the use and disposal of PCB-containing items.
- Retired and inactive mine lands: Abandoned and dormant mines may not even have been thoroughly cleaned up and may contain characteristics varying from exploration excavations to full-scale, large-scale mine shafts, pits, waste dumps, and processing areas.

5.4 MICROPLASTICS

As per NOAA and the European Chemicals Agency, microplastics are remnants of any sort of plastic that are smaller than 5 mm in length. They pollute natural environments by entering through a range of origins, such as cosmetics, textiles, packaging materials, and industrial activities (Mekuria et al., 2009).

Microplastics are currently classified into two types. Primary microplastics are defined as any plastic pieces or particles that are 5.0 mm or smaller before infiltrating the ecosystem. Microfibers from apparel, microfibers, and plastic pellets or nurdles are examples. Secondary microplastics are formed when bigger plastic goods degrade in the environment due to natural weathering processes. Water and drink bottles, fishing equipment, plastic bags, microwave dishes, tea bags, and tire wear are also potential sources of secondary micro - plastics.

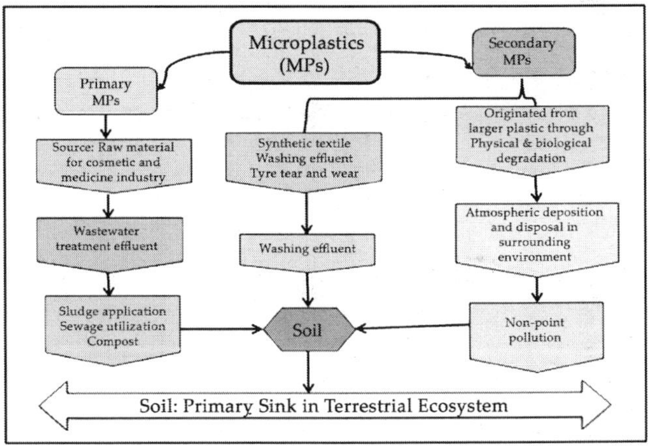

Figure 5.9: Effects of microplastics in the soil environment.

Source: Kumari A, Rajput VD, Mandzhieva SS, Rajput S, Minkina T, Kaur R, Sushkova S, Kumari P, Ranjan A, Kalinitchenko VP, Glinushkin AP. Microplastic Pollution: An Emerging Threat to Terrestrial Plants and Insights into Its Remediation Strategies. Plants. 2022; 11(3):340. https://doi.org/10.3390/plants11030340

Both varieties are known to endure at significant levels in the ecosystem, especially in marine and aquatic habitats, wherein they produce water contamination. Textiles and clothing account for 35 percent of all marine microplastics, mostly due to the evaporation of polyester, acrylic, or nylon-based clothes, which occurs frequently during the washing process. Microplastics, on the other hand, concentrate in the atmosphere and terrestrial environments. Microplastics are distinguished from bigger plastic trash, like plastic bottles, by the phrase macroplastics.

Because plastics decompose slowly, microplastics are highly likely to be ingested, incorporated into, and accumulated in the bodies and cells of a wide range of creatures. Toxic chemicals from the sea and drainage can potentially biomagnify their way up the food chain (Meshesha et al., 2012). Microplastics have been shown in terrestrial ecosystems to impair the health of the soil environment and the mass of earthworms. The cycle and transportation of microplastic pollution are not entirely known, although research into the subject is now happening. Deep layer marine sedimentation investigations in China confirmed the existence of plastics in accumulation layers that predate the creation of plastics, implying that microplastics in top sampling ocean studies are likely underestimated. Microplastics were also discovered in the mountain ranges, far from their origins.

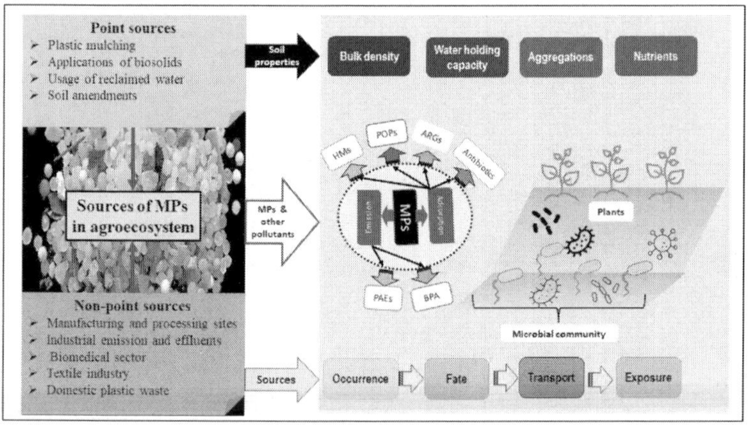

Figure 5.10: Development and application of tests for microplastic detection in soil.

Source: Kumari A, Rajput VD, Mandzhieva SS, Rajput S, Minkina T, Kaur R, Sushkova S, Kumari P, Ranjan A, Kalinitchenko VP, Glinushkin AP. Microplastic Pollution: An Emerging Threat to Terrestrial Plants and Insights into Its Remediation Strategies. Plants. 2022; 11(3):340. https://doi.org/10.3390/plants11030340

Although a significant fraction of microplastics is projected to wind up in the world's soil, little to no studies on micro - plastics in soil beyond the marine contexts have been undertaken. Microplastic concentrations have been observed to have a negative connection with vegetation coverage and root density in wetland habitats. There is some possibility that filamentous secondary microplastics from household appliances may reach the soil due to water processing plants' inability to totally filter out all of the micro - plastic fibers (Ma, Zhao, & Zhu, 2021).

Moreover, geophagous soil fauna including earthworms, mites, and collembolans may influence the quantity of secondary microplastic in soil by turning eaten plastic trash into microplastic through digestive mechanisms. Nevertheless, more research is required. There is concrete evidence tying the utilization of organic waste materials to the existence of synthetic fibers in soil; nonetheless, most studies on polymers in soil only record their presence without mentioning their origin or quantity. Controlled experiments on fiber-containing land-applied wastewater sludges deposited in soil revealed validated fiber recovery a number of years later.

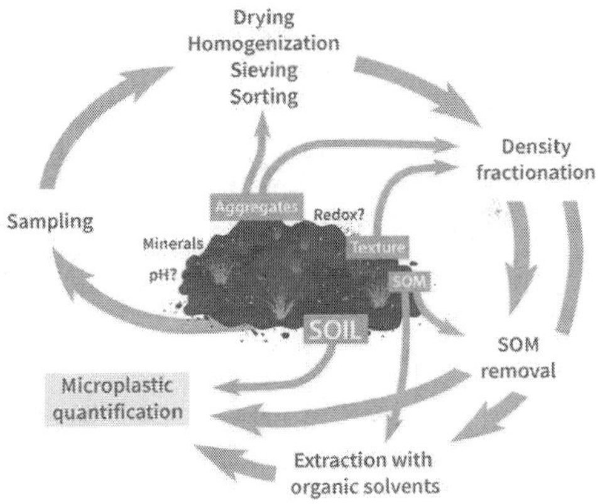

Figure 5.11: Despite the fact that the majority of plastic pollution originates on land, current study is mostly focused on aquatic habitats.

Source: Thomas D, Schütze B, Heinze WM, Steinmetz Z. Sample Preparation Techniques for the Analysis of Microplastics in Soil—A Review. Sustainability. 2020; 12(21):9074. https://doi.org/10.3390/su12219074

5.5 HYDROCARBONS

Petroleum is primarily made up of numerous hydrocarbon complicated structures. Chromatography can separate it into saturated hydrocarbon, aromatic hydrocarbon, and non-hydrocarbon molecules. The saturated hydrocarbon molecular composition is made up of carbon-carbon bonds and carbon-hydrogen bonds, both of which are easily destroyed. Furthermore, because its boiling point is relatively low, the saturated hydrocarbon can progressively evaporate from the soil by photosynthesis and volatilization. The molecular composition of aromatic hydrocarbons, on the other hand, is highly complex. The intricate benzene ring and its increased boiling point make removal from the soil much more difficult. The polycyclic aromatic hydrocarbon (PAH) is a common POP found in a variety of environmental systems, and the EPA of the United States and the European Community have placed PAHs on their lists of prioritized control contaminants.

Non-hydrocarbon substances have the highest carbon number of the three types of hydrocarbons. Because they are water insoluble and have a high fusing and boiling point, they are hard to separate from soil, culminating in clear environmental contamination and genotoxicity. As a result, the organic soil leftover toxicity is primarily composed of PAH and nonhydrocarbon chemicals. Furthermore, non-hydrocarbon makeup includes a trace quantity of heavy metal components like vanadium and nickel, and some research teams have found that drilling fluid includes heavy metal contaminants with toxicological properties, durability, and refractory properties (Lim et al., 2020).

Aside from oil, wastewater in oil and gas operations can pollute the land. At the moment, most national oil resources have reached the middle to late stages of oil output, with water content in crude oil reaching 70%, 80%, and up to 90% in some cases. Following the separation of oil and water, a large amount of oily sewage will be produced. They could cause major soil and water contamination if left unchecked. Oilfield sewage comprises oil, different salts, organic stuff, inorganic matter, and certain bacteria, among other things. This type of waste water not only promotes soil salinization, but it also degrades the integrity of the soil ecosystem.

Figure 5.12: This photo shows oil floating on water in contaminated soils south of New Orleans.

Source: https://phys.org/news/2010-06-amount-oil-contaminated-soils.html

According to a wide number of studies, the damage of oil-polluted soil mostly involves the following facets: Petroleum is rich in carbon and a small number of nitrogen compounds, so it can modify the physicochemical characteristics of soil organic matter and affect the C/N, C/P, salinity, pH, EH, and conductivity of soil. Heavy metals (nickel and vanadium) in oil mixes, as well as high salt concentrations in oilfield output water, can potentially harm the soil ecology.

Second, microorganisms in the organic ecosystem are plentiful in clean and healthy soil. In a normal condition, microorganisms that can withstand the stress of oil pollution do not develop; but, in polluted soil, in an attempt to acclimatize to this sort of habitat, they can build specific enzyme systems and progressively form a dominant population with a symbiotic or synergistic impact. A handful of investigations have indicated that hydrocarbon contamination can alter the microbial population, the makeup of the ecosystems, and the enzymatic activity in soil, with the inhibition effect taking precedence (Li et al., 2018).

Third, it can hamper normal crop growth by lowering germination rates and quality, as well as decreasing tolerance to pests and pathogens. Furthermore, the oil molecules may react with exogenous nitrogen and phosphorus, restricting nitrification and the elimination of phosphoric acid, lowering the functional nitrogen and phosphorus in the soil and affecting crop absorption. Furthermore, the pahs found in petroleum chemicals are carcinogenic, mutagenic, teratogenic, and poisonous. It can permeate people's and animals' bodies by inhalation, skin contact, and eating, deteriorating the normal operation of livers and kidneys, among other things, posing a serious threat to human health.

Finally, oil contaminants in the soil have an effect not only on the pedosphere, but also on the air and the aquatic sphere. To be more precise, low boiling point and light weight hydrocarbons can effortlessly infiltrate the atmosphere by evaporation; next penetrate the groundwater system via runoff and penetration into surface water and osmosis; and lastly reach human bodies via the food chain (Meli, 2013).

Several nations across the globe have done numerous studies on the repair of oil-polluted soil in recent times. Physical, chemical, microbiology, and plant remediation are the most commonly used procedures. Traditional physical and chemistry approaches, such as soil extraction and replacement, elution, heat treatment and thermal resolution, extraction-separation, and chemical oxidation, are utilized to manage physical mechanisms or chemical reactions in an attempt to transform the physical properties of the soil and properly handle contaminants (García-Ruiz et al., 2017). This strategy is usually more complete and steady, but it does necessitate the construction of some fixed treatment structures, the employment of chemical agents, and the management of temperature, pressure, and power supply to guide the entire operation. It is rarely commonly employed in actual application due to its high cost, secondary contamination, and inability to remediate broad areas.

Figure 5.13: Soil contaminated by an oil spill: an example off the coast of Mauritius (Indian Ocean).

Source: Seveso D, Louis YD, Montano S, Galli P, Saliu F. The Mauritius Oil Spill: What's Next? Pollutants. 2021; 1(1):18-28. https://doi.org/10.3390/pollutants1010003

In a nutshell, restoring petroleum-polluted soil is a difficult task. At the moment, physical and chemical restoration procedures mostly rely on heat treatment and chemical dissolution methods, which can swiftly treat large-scale contaminated soils. While certain other low-cost, no-secondary-pollution treatment methods also need to be designed, particularly for rapid emergency service, like well blowouts, oil leakage accidents, etc.

Now, bioremediation technology is making significant advances. It was discovered through extensive literature research that indigenous microorganisms can maintain high interaction and degradation performance, but there are too few in the natural ecosystem to fulfill the actual requisites; thus, how to isolate, sift, and develop more effectual in-situ biodegradable microscopic organisms is an access point (Guerra et al., 2017).

In terms of phytoremediation, most contaminants are removed mostly through the breakdown of rhizobacteria or conversion, so creating an effective soil plant rhizosphere microorganism regeneration process is critical. In actuality, the makeup of petroleum contaminants is significantly more complicated; the majority of soil residues contain high-molecular-weight PAHs and non-hydrocarbon molecules, which may be followed by soil salinization and heavy metal contamination.

Provided this, it is complicated to fix the issue completely with a single intervention project; consequently, looking for an extremely effective and reliable microbes-plant-physiochemical joint repair mechanism that is best suited for the field survey and how to transform the technology into non-point source contamination repair has quite promising application possibilities.

Figure 5.14: Kelp after an oil spill.

Source: https://en.wikipedia.org/wiki/Oil_spill#Environmental_effects

5.6 ELECTRONIC WASTE

The widespread use of digital gadgets such as mobile phones, tablet devices, computers, and wearable technology has created a new ecological hazard.

Discarded electronics generate vast heaps of e-waste, which pollutes soil and groundwater in the Indian capital, according to a research article.

The study's findings suggest that soil and water pollution is certain to enter the human body, posing health risks. Krishna Vihar, in Delhi, is home to unorganized e-waste handling and disposal units, as well as recycling facilities that discharge toxic effluents, notably heavy metals, straight into open spaces. Solid garbage is frequently disposed of simply by igniting it in the outdoors(Ha, Van Dung & Ngoc, 2013).

Soil and groundwater measurements were obtained from five locations chosen for their proximity to e-waste operations. To evaluate their results, the researchers chose five additional localities with no e-waste-related activities. The investigation discovered hazardous levels of pollution in soil and groundwater near uncontrolled and badly contaminated e-waste landfills.

The average levels of heavy metals (copper, lead, cadmium, and chromium) in topsoil samples from e-waste sites was found to be much above the range observed in typical agricultural soil. When contrasted to standard site top soil samples, the mean copper level in e-waste site top soils was roughly 30 times higher. The copper content in top soil was determined to be 283.23 mg/kg. Lead came in second with a value of 298.10 mg/kg (Haregeweyn et al., 2015).

In comparison to reference site samples, which had insignificant cadmium amounts, the average cadmium level in top soils was 16 times higher than that of agricultural norms. This could imply that cadmium and related compounds, which are commonly used in electrical and electronic equipment, have made their ways into the soil in locations with substantial e-waste recycling activities.

Toxic compounds will be discharged into the environment if e-waste is not managed in a sustainable manner. When exposed to the environment, cadmium, chromium, lead, and other heavy metals leak into groundwater. It quickly enters the food chain via bio-accumulation and affects your liver and kidney.

Due to the use of pseudoscientific recycling technologies, heavy metal concentrations in soil are high in regions where e-waste modules are situated. Copper, for example, leaches into the soil after being retrieved from printed wiring boards of electronic devices due to incorrect e-waste deconstruction and treatment. Slowly, the metal accumulates on the surface of polluted soils, with almost no downward movement.

The concentrations of all heavy metals, save zinc, in water near e-waste treatment plants were also higher than the permissible limits. The average proportion of copper in water samples from the e-waste site was found to be 30 times higher than water limits and eight times higher than benchmark water levels. The copper concentration in the e-waste site water collection was discovered to be 1.465 mg/litre (Ighodaro, Lategan, & Yusuf, 2013).

When the investigation began in 2012, the region was dominated by live recycling workshops. The extraction of rare metals from e-waste proved dangerous. Acid treatment was used to recover metals such as copper, leaving deposits in acid containers. These discharges were then dumped directly into the ground.

5.6.1 Effects

Contaminated lands can present a number of health and environmental risks. Since this degree of contamination is minimal and the danger of exposure to toxic or hazardous substances is low, certain contaminated sites present minimal harm to human health and the environment.

Other contaminated locations are of more significance due to the chemicals that could be contained and their proclivity to stay in or travel through the ecosystem, providing threats to humans or the environment. These sites have to be carefully maintained through confinement or remediation to avoid hazardous contaminants from harming humans, wildlife, or natural systems both on and offsite.

- Contaminated soils can leak harmful chemicals into surrounding ground or surface waters, where they can be consumed by plants and animals, taint a human drinking water source, or volatilize and poison the indoor air in neighboring houses.
- In dry locations, pollution in soil can be spread further by wind-borne dust. When soil contamination enters streams, it can collect in sediments, which can be challenging to treat and have an impact on local ecology and human health (Issaka& Ashraf, 2017).
- Contact with poisonous and toxic substances on a polluted area can cause harm to humans through exposure to polluted soil, air, surface water, and groundwater.
- Humans and wildlife can be subjected to toxins by inhalation, ingestion, or skin contact when polluted soils are not appropriately managed. Human exposure dangers vary by location and are difficult to extrapolate at the state scale. Acute or chronic consequences are possible.

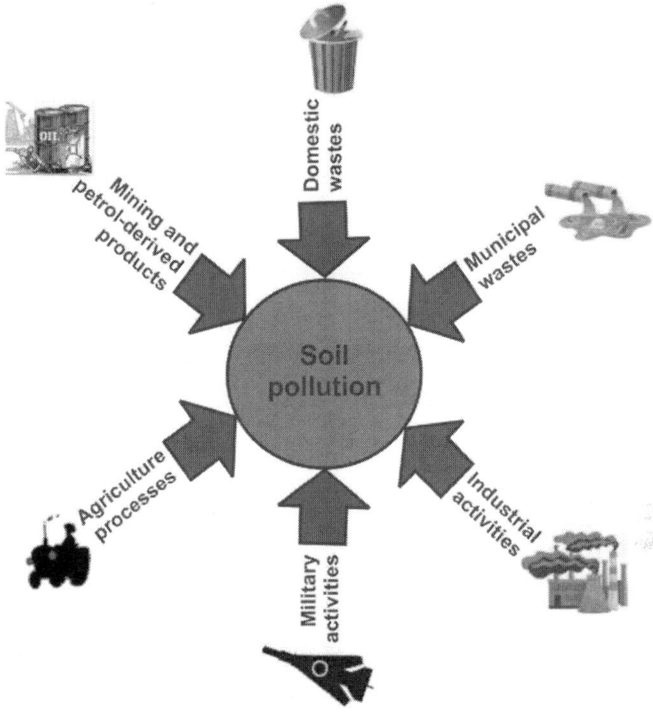

Figure 5.15: The health risk associated with contaminated soil will depend on the type of hazardous substance, the exposure route and the length of time a person is exposed.

Source: Belviso C. Zeolite for Potential Toxic Metal Uptake from Contaminated Soil: A Brief Review. Processes. 2020; 8(7):820. https://doi.org/10.3390/pr8070820

5.7 SITES OF SPECIAL CONCERN

Industrial and manufacturing sites

Soils at industrial and manufacturing locations are frequently polluted by a variety of toxins. The sort of contamination will be determined by the factory's output. Pollution can arise when chemicals spill from structures or vehicles onto the soil. At times, the plant may have a disposal heap or holding area that was once thought to be safe but is now proven to be a source of pollution. Industrial sites can be extremely huge as well. As a result, full-site soil restoration is a costly and complex, but vital, undertaking (Jomaa et al., 2013).

Landfills, junkyards and waste disposal sites

Landfills, junkyards, and garbage disposal facilities, like industrial facilities, reflect a significant potential for soil contamination. These locations frequently contain a diverse range of contaminants such as lead, arsenic, and oil products. All of them are hazardous to human welfare on their own. When they interact, they may condense to produce even more hazardous chemicals. Isolation and decontamination of these places are expensive, technically difficult, and logistically difficult.

Highway corridors, parking lots, areas of heavy traffic

Vehicles are omnipresent in the urban environment. Thousands of automobiles and trucks are strewn across highway channels, squeezed into parking areas, and crammed into major roadways with heavy traffic. Our automobiles transport us from one point to the other; we are frequently reliant on them. Though they may aid in transportation, places with significant numbers of automobiles face a number of pollutant threats due to emissions and fluids leaking. Lead, for instance, can be found in high concentrations in places with high traffic pollution, and petroleum or oil spills on roads or parking spaces can be swept into neighboring soils.

Household sites

Households might not be apparent sources of soil contamination. However, soil contamination can occur during the construction of a home. Leaks of hydrocarbons from heavy machinery are possible. Paint may splatter. Fibers from roofing materials can blow downwards and harm soil life. These are only a few instances of soil pollution on building sites (Kavian et al., 2019).

Furthermore, homeowners may unintentionally poison their soil. The principal issue is the over-application of pesticides or herbicides. Chemically processed wood is occasionally used in beautification. If not utilized correctly, this can pollute the soil as much as the plants and microbes that live in it. Over use of winter salt can also be detrimental to your soils.

Farmlands with build-up of contaminants

In the first half of the 20th century, numerous insecticides in the United States were made with lead-arsenate. At the time, experts and farmers were unaware that lead posed a health risk. As a consequence, lead can still be discovered in the soil of abandoned farms today. Furthermore, herbicides have been extensively developed and manufactured since the 1950s. These chemicals must be handled correctly; inappropriate application can damage the soil, plants, and even human health. Depending on the crop and fertilizer type utilized, high-load fertilizer treatments may generate pollution in soils.

5.8 EXPOSURE TO CONTAMINANT

5.8.1 Ingesting Soil

Though it may appear strange to eat soil, pollutants can be presented in a variety of circumstances. Youngsters may be especially vulnerable since they play in raw earth. Dust particles that typically scatter during play may be inhaled by children. They could also be inquisitive and attempt to eat soil. If the soil is not poisoned, this has no effect on the children. For a sense of security, you can have the soil in your backyard analyzed for lead and other contaminants.

Contaminated soil particles could also have an impact on our food production. Contaminated soil, for instance, could be found on vegetables.

If an item, such as lettuce, is grown in contaminated soil, the leaves may get contaminated. Washing lettuce is critical. In the store, root crops like potatoes and carrots generally have soil on them. If they came from hazardous soil, they must also be thoroughly washed (Kumarasinghe, 2021).

When soil is left naked, the risk of ingesting it is greatest. Covering the soil with grasses or other plants and covering it thoroughly decreases the chance of contamination. When individuals eat outside near windswept soil on a windswept day, airborne pollutants may land on their food before they eat it.

5.8.2. Breathing Volatiles and Dust

Fine particles might become airborne if soils are exposed due to wind or other disturbances. Soil dust can be produced by construction or demolition activities, mining operations, or bad landscaping attempts. Inhaling polluted dust can cause physical and chemical harm to individuals. Asbestos fibers, for example, can penetrate the lungs. Lead, for example, can harm the neurological system, including the brain.

5.8.3. Absorbing through Skin

Contaminants can also enter the body through the skin. In the United States, creosote is a common substance used to protect timber. This complicated chemical concoction can seep out of wood material and pollute the soil. If creosote-contaminated soils are handled, the skin may burn, peel, or become severely reddened over time (Lal, 2019).

5.8.4. Eating Food Grown in Contaminated Soil

Whenever you put in a lot of effort in your personal or community gardening, be certain the soil is free of contaminants! There is a possibility that if you produce vegetables in polluted soil, your product will be tainted as well.

Some housing projects and community gardens have been developed in locations that previously served a different purpose. Pollutants may be found if the land was an industrial or manufacturing area or the site of a destroyed structure. There are methods for "fixing" polluted soil, but you ought to analyze it first.

Many plants and herbs have the ability to absorb pollutants as they develop. If you eat them, you are putting yourself in danger. Vegetables and plants may also have soil particles on them.

Contaminants remain if adequate washing is not performed. Some garden setups could also have chemically processed wood linings. It is advisable to examine your soil if you did not construct your flower beds manually because contaminants can leech into the garden soil. Lastly, gardens or farms might well be built on clean soils adjacent to a polluted area. In this situation, dust from the polluted location may blow into the previously untainted soil, polluting it. In these circumstances, it's best to be aware of how your neighbor's land is being used.

5.9 PESTICIDES AND HERBICIDES

Pesticides include a large group of substances. Organochlorine (OC) insecticides, which were used effectively to manage a variety of illnesses like malaria and typhus, were prohibited or limited in most technologically sophisticated states after the 60s. The emergence of various synthetic insecticides, such as organophosphate (OP) insecticides in the 1960s, carbamates in the 1970s, and pyrethroids in the 1980s, as well as herbicides and fungicides, considerably improved pest control and agricultural productivity. A pesticide should ideally be fatal to the intended pests but not to non-target species, especially humans. Sadly, this is not the reality, so the debate over pesticide use as well as abuse has erupted. The widespread usage of these compounds, based on the idea "if a little is helpful, a lot more would be great," has wreaked havoc on humans and other living things (Lefèvre et al., 2021).

The principal benefits are the side effects of pesticides – the direct gains anticipated from their employment. For instance, eliminating caterpillars that eat on the crop has the additional advantages of increasing yields and improving vegetable quality. The three main outcomes result in nearly thirty primary advantages spanning from recreational grass preservation to human life saved. Secondary gains are those that are less apparent or visible as a byproduct of the principal benefits. They can be subtle, less visible at first glance, or long-term. As a result, secondary advantages are more difficult to demonstrate correlation and causation for, yet they can still be persuasive justifications for pesticide application.

For instance, a larger cabbage production may generate additional cash that can be used to fund public learning or medical treatment, resulting in a stronger, more literate population. Secondary advantages have been observed, ranging from healthier people to saved biodiversity.

Pesticides have the potential to pollute soil, water, turf, and other flora. Pesticides, in addition to destroying insects and weeds, can be hazardous to a variety of other creatures such as birds, fish, helpful insects, and non-target vegetation. Although insecticides are the most severely hazardous type of pesticides, herbicides can also be harmful to non-target organisms.

A vast amount of transformation products (TPs) from various pesticides have been identified. Not all probable pesticide TPs have been detected in soil, indicating that more research in this area is urgently needed. Some characteristics, including water solubility, soil-sorption constant (Koc), octanol and water partition coefficient (Kow), and half-life in soil, influence the persistence and mobility of these pesticides and their TPs (DT50). Pesticides and TPs can be classified into two types: (a) hydrophobic, persistent, and bioaccumulable pesticides that are highly bonded to soil. DDT, endosulfan, endrin, heptachlor, lindane, and their TPs are examples of pesticides that display this tendency. The majority of them are now prohibited in agriculture, although their residues remain (Kheyrodin, 2016).

(b) Polar pesticides primarily consist of herbicides, although they also include carbamates, fungicides, and several organophosphorus insecticide TPs. They can be transferred from the soil through runoff and leaching, posing a hazard for the population's water supply. Herbicide pesticide TPs are arguably the most studied pesticide TPs in soil.

Several metabolic processes have been proposed, using hydrolysis, methylation, and ring breakage to yield a variety of hazardous phenolic chemicals. Pesticides and their TPs are maintained to varying degrees by soils, based on the interaction among soil and pesticide characteristics. The organic material concentration of the soil is the most influential feature. The higher the organic matter concentration, the more pesticides and TPs are absorbed.

With paraquat and other positively charged herbicides, the soil's ability to store positively charged ions in an interchangeable state is critical. Mineral acids are required for extraction, and no analytical advancements or studies have been recorded in recent years. The pH of the soil is also important. For ionizable insecticides, adsorption increases as soil pH decreases.

Pesticide use on a large scale can lead to a drop in the population of beneficial soil microorganisms. According to the soil scientist, if both bacteria and fungus are lost, the soil will disintegrate. Abuse of agrochemicals has comparable consequences on soil organisms as human misuse of antibiotics (Kelly et al., 2020).

Chemicals used indiscriminately may work for a few years, there are just not enough favorable soil microorganisms to hang onto the micronutrients after a little while. Plants, for instance, rely on a range of soil microbes to convert nitrogen from the atmosphere into nitrates that plants may use.

Prevalent landscape herbicides destabilize this operation: triclopyr inhibits soil bacteria that convert ammonia to nitrite; glyphosate restricts the production and interaction of free-living nitrogen-fixing bacteria in soil; and 2,4-D reduces nitrogen fixation by bacteria living on the root system of bean plants, decrease the growth and the development of nitrogen-fixing blue-green algae, and impedes soil bacteria that convert ammonia to nitrates. Many plants have mycorrhizal fungi that grow alongside their roots and assist in nutrient absorption.

Herbicides in the soil can also harm these fungi. According to one investigation, oryzalin and trifluralin both suppressed the development of specific mycorrhizal fungi. Roundup has been proven in laboratory trials to be toxic to mycorrhizal fungi, with some detrimental effects observed at doses lower than those observed in soil after usual applications. Triclopyr was also discovered to be harmful to some mycorrhizal fungi species, and oxadiazon lowered the amount of mycorrhizal fungal spores.

5.10 CONCLUSION

Soil contamination is caused by a variety of human activities and experiments that end up poisoning the soil. The most prevalent sources of soil contamination are industrial wastes such as hazardous gases and chemicals, as well as agricultural pesticides, fertilizers, and insecticides. Others include a lack of knowledge about soil management and associated systems, unfavorable and dangerous irrigation methods, inappropriate septic system management and maintenance, and sanitary sewage leaks. There is an immediate need for a graded method to assess the ecological danger of contaminated soils. As a first rung, generic soil testing levels are required. Higher levels of ecological risk assessment, on the other hand, should include some type of site-specific evaluation (Keesstra et al., 2018).

It is also critical to organize the numerous research inside a structure or management support system that is clear and relevant to all stakeholders. To address these ambiguities, a strength of the evidentiary method may be an immediate answer. The TRIAD method, which combines and classifies data in a triangle - chemistry, toxicology, and ecology – is a useful technique for dealing with conceptual difficulties. Several solutions to these flaws have been presented. In terms of ecotoxicity, direct testing could permit significant improvements in risk assessments. In terms of human health hazards, integrating biological accessibility in risk calculations, making better use of current knowledge regarding exposure routes, dose-effect relationships, and combination impacts, and biomonitoring of consequences are all possibilities for advancement.

CHAPTER 6

HOW ARE PEOPLE EXPOSED TO SOIL CONTAMINANTS?

CONTENTS

6.1 Ingesting Dust or Soil Directly 140

6.2 Ingesting Food or Vegetables Grown in Contaminated Soil or with Foods in Contact with Contaminants 146

6.3 Skin Contact with Dust or Soil 155

6.4 Vapors from the Soil 158

6.5 Inhaling Dusts While Working in Soil or Windy Environments 162

6.6 Exposure to Contaminated Ground Water 164

6.1 INGESTING DUST OR SOIL DIRECTLY

Among the ways through which contaminated soil finds its way to the body of human beings is through ingesting dust or consuming dust directly. Ingestion of dust occurs when small soil particles become air borne as a result of wind or any other form of disturbance. There are various ways through which contaminated soil particles are released into the atmosphere. They include construction or demolition work, poor landscaping or mining activities. By disturbing contaminated soil particles of contaminated soil are released into the atmosphere where it is inhaled various kinds of damage to the body. When soil is exposed to various chemicals such as acids, the same chemicals find their way into the brain after inhalation. This may damage the nervous system in various ways (Zhang & Huang, 2015). Natural events such as strong winds and storms increase the chances of inhaling contaminated soil particles. As children play in contaminated soil, they can breathe in dust particles that are dispersed during playing activities. Dust can also find its way into people's food when left outside on a windy day or in windy soil. When blown, soil particles enter foods which is the consumed causing various drastic effects. Dust inside buildings could also lead to contamination. In most cases, dust particles tend to settle on objects or surfaces and can easily be ingested by both children and adults.

Figure 6.1: Children are most likely to ingest soil as they play.

Source: https://www.123rf.com/photo_98773374_cute-children-on-play-ground.html

Exposure to soil contaminants can also occur through direct ingestion of contaminated soil. Children are more susceptible to this kind of exposure as they have a great tendency to play in soil. They are also very curious- a trait that may cause them to try and eat soil. If the soil sample is contaminated, the child may suffer from the effects of the contaminants. In most cases, yards are usually tested to avoid any kind of contamination. Contamination can also occur through soil dust contamination on food supplies. If food substances are not properly cleaned then the contaminated soil particles will be present in food (Yakupoglu, Rodrigo-Comino &Cerdà, 2019). For food items such as lettuce and kale that grow in contaminated soils, their leaves may be covered by contaminated soil particles. For this reason, it is very important to ensure that the leaves are properly cleaned. When harvesting beans locally, the method of extraction provides several opportunities for beans coming to contact with contaminated soil particles. When not properly cleaned, these soil particles will be ingested. Crops such as potatoes and carrots usually have soil in them when stored. Proper cleaning is done on such root crops.

Figure 6.2: Food supplies such as beans and carrots can be contaminated when exposed to contaminated soils.

Source: Luby CH, Goldman IL. Improving Freedom to Operate in Carrot Breeding through the Development of Eight Open Source Composite Populations of Carrot (Daucus carota L. var. sativus). Sustainability. 2016; 8(5):479. https://doi.org/10.3390/su8050479

Exposure to contaminated soil and dust is a common route for children and adults more so children. Studies show that children have a higher likelihood of exposure as they play on floors indoors and outdoors. Exposure is further increased by their tendency to mouth their hands, toys and other objects. Some children are exposed to contaminated soil through deliberate hand-to-mouth movements, mouthing of objects that drooped on the floor or through unintentional eating of food. Other than children, pregnant women are very susceptible to contamination through direct soil contamination and dust (Webb, Herrick, &Duniway, 2014). Ingestion of soil contaminants directly or through dust brings about the term soil pica. Soil pica is used in understanding and estimating general exposures to environmental chemicals. Pica is a concept explaining the ingestion of non-food substances or ingesting large amounts of certain foods. With regards to ingestion of soil contaminants, pica can be used to refer to recurring or repeated ingestion of soil substances. For this reason, soil pica is defined as the ingestion of materials that are defined as flower-pot soil, yard soil and clays all classified as soil.

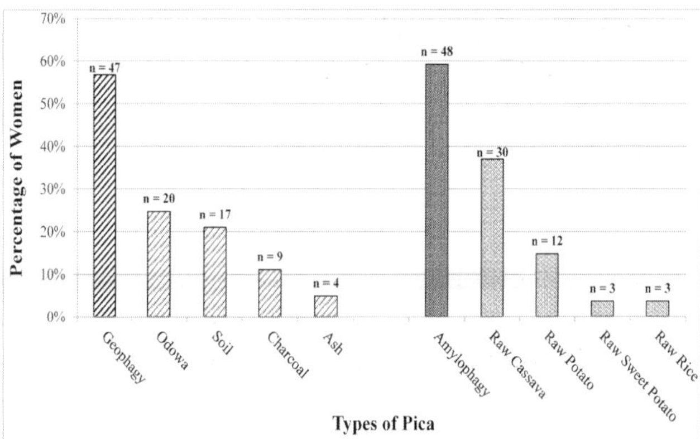

Figure 6.3: Soil pica is the ingestion of soil.

Source: Chung EO, Mattah B, Hickey MD, Salmen CR, Milner EM, Bukusi EA, Brashares JS, Young SL, Fernald LCH, Fiorella KJ. Characteristics of Pica Behavior among Mothers around Lake Victoria, Kenya: A Cross-Sectional Study. International Journal of Environmental Research and Public Health. 2019; 16(14):2510. https://doi.org/10.3390/ijerph16142510

Though soil pica is very common, there is little information about it. The little available information shows that there is a variation of soil pica across age, race and income level. This behavioral tendency is mostly noted among children less than years. Among adults, soil pica is very rare though geophagy is rare among children. There is a very distinct difference between geophagy and soil pica with regard to the implications on public health. As geophagy is the consumption of soil such as clay from uncontaminated sources, there may be no effects of consumption. Soil pica mostly involves the consumption of surface soils in most cases the top 2-3 inches. There are different reasons or motivations for geophagy or human soil pica behavior. They include gastrointestinal distress, nutritional deficiencies and alleviated hunger. Other motivations include the desire to remove toxins from the body, cultural influences, physiological influences and the urge to self-medicate.

In most cases, consumption of contaminated soil through direct ingestion or as dust is due to hand-to-mouth behavior. For this reason, soil properties related to adherence to skin may prove vital. Some of these properties include moisture content, organic matter content and soil particle size. The extent to which soil particles adhere to skin makes soil available for ingestion. There are different ranges of soil particles. For instance, sand soil is 50-2000 micrometers, clay is less than 2 micrometers, and silt is 2-50 micrometers. This attribute makes soil easy to ingest. Dust particles in the atmosphere have a size range of 0.001 to 30 micrometers. Studies on ingestion of soil particles show that fine soil particles tend to have an efficient adherence to human hands. Studies also show that soil fractions are independent of soil origin and organic matter content. Soil moisture content is greatly linked to skin adherence in that high moisture contents tend to adhere to the skin. Studies show that about 90% of soil particles in playgrounds 100 micrometers in size tend to adhere to children's hands (Van Leeuwen et al., 2019). The levels of contaminants found in soil differ according to soil particles. When evaluating exposure to contaminants through ingestion of soil can be done through the evaluation of indoor settle dust.

Figure 6.4: People who handle soil regularly are likely to carry contaminated soil.

Source: https://www.frontiersin.org/research-topics/36919/sustainable-soil-fertility-practices-for-smallholder-farmers

Exposure to contaminants to soil dust can either be done through indoor settle dust or outdoor settled dust. There are soil particles found on the earth's surface as unconsolidated mineral or organic matter and are mostly found outdoors. In some cases, they are found indoors when plants are grown indoors. Dust particles can settle on outdoor objects or surfaces. Indoor settled dust particles are those found in the interior sections that may have settled on carpeting, floors, surfaces or objects. These particles may have found their way into the building after being blown into the indoor environment from outdoors. During this time, organic matter may also find its way into the house. Outdoor settled dust particles are those that have settled on outdoor surfaces and objects. This may be as a result of dry or wet deposition. It may be quite difficult to distinguish between outdoor and soil that has settled indoors. This may be due to the fact that dust found on surfaces is the layer of soil.

Results from various studies have been very useful in evaluating high soil contact scenarios. Among the factors influencing soil consumption in traditional areas in the lifestyles of people living in those areas. Studies show a high rate of soil ingestion among individuals from those groups. This makes these groups of people highly likely to suffer from contamination through soil ingestion. Among these groups of people, ingestion rates in adults is about 50 mg/ day with the upper percentile being about 200 mg/per day. Continuous consumption of contaminated soils means that there will be a build-up of the toxic material in their bodies making the effects more harmful. In some cases, the upper percentile value can be achieved making the risk of exposure very high. When the amount of soil consumed is about 45% while dust is about 55%, this means that the upper percentile soil values would be 20 mg/day while that of dust is 90 mg/day. When calculating using upper percentile values then the values would be 30 mg/ day and 100 mg/ day respectively (Tsoraeva et al., 2021).

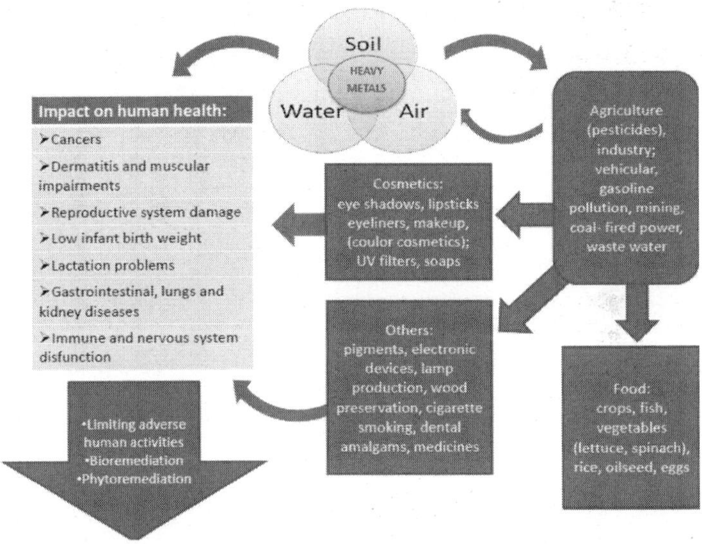

Figure 6.5: Continuous consumption of contaminated soils leads to build up of contaminants in the body.

Source: Witkowska D, Słowik J, Chilicka K. Heavy Metals and Human Health: Possible Exposure Pathways and the Competition for Protein Binding Sites. Molecules. 2021; 26(19):6060. https://doi.org/10.3390/molecules26196060

For most pregnant women, consumption of soil is mostly attributed to their cravings and aversions of either food or non-food items. These urges are mostly developed during their pregnancy. They may also tend to consume different kinds of beverages. Some of the non-food items they consume during their pregnancy include ice and chalk from river clay banks. By ingesting clay from river clay banks, they can consume contaminated soil samples. In some parts of the world, there are women who eat certain stones during their pregnancy. If contaminated, women are exposed to the contaminants through direct ingestion (Tully et al., 2015). Exposure to soil is mostly noted when soil is left bare. Risks of contamination can be reduced by covering soil with grass, mulch or other plants. When holding outdoor events or parties on a windy day, there is a need for proper measures to be put in place to ensure that contaminants do not land on food. Various measures are also put in place to ensure that there are fewer dust particles on objects and surfaces. This may include regular cleaning and dusting of household equipment.

6.2 INGESTING FOOD OR VEGETABLES GROWN IN CONTAMINATED SOIL OR WITH FOODS IN CONTACT WITH CONTAMINANTS

Exposure to soil contaminants is mostly through ingestion of vegetable products grown in soils filled with soil contaminants. There are cases where individuals grow products in their gardens with no knowledge of whether the soil they are using has been contaminated. There are high risks of contamination when food is grown in contaminated soil. There are various ways soil can be contaminated. For instance, there are housing developments and community gardens that have been set up in areas that served a different purpose historically. Contaminants may be present in sites where the area was a manufacturing or industrial site or an area where a building was demolished. For this reason, gardening soil should be tested and fixed if possible. Exposure to contaminated food occurs when both herbs and vegetables absorb contaminants as they grow placing great risks on individuals as they eat.

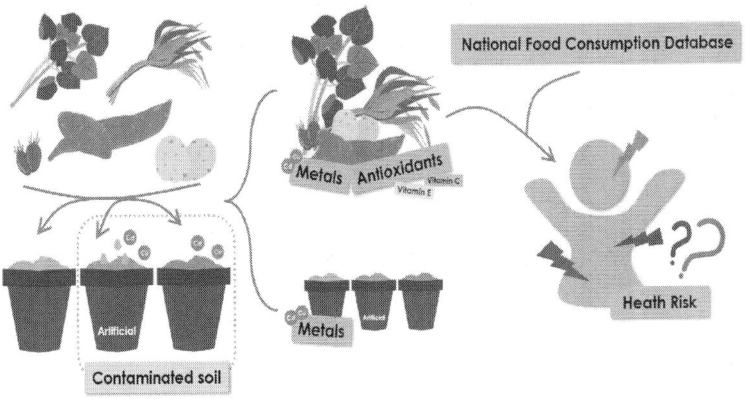

Figure 6.6: Vegetables growing in contaminated areas are likely to absorb some of the contaminants.

Source: Huang W-L, Chang W-H, Cheng S-F, Li H-Y, Chen H-L. Potential Risk of Consuming Vegetables Planted in Soil with Copper and Cadmium and the Influence on Vegetable Antioxidant Activity. Applied Sciences. 2021; 11(9):3761. https://doi.org/10.3390/app11093761

One way through which contamination through the ingestion of vegetables occurs is when contaminated soil dust is found on the surface of plants. If not properly cleaned, the particles will be eaten and accumulate in the body. There are cases where garden beds are built and lined with chemically-treated wood. In most cases, garden owners will build the gardens themselves while ensuring that no chemicals leach into garden soil. Sites for locations of gardens are really important (Telles, Guimarães, &Dechen, 2011). There are farms located next to areas of contamination. During tilling activities, the contaminated soil may find its way into uncontaminated soil. Strong winds can also cause the movement of contaminated soil to uncontaminated soils. Soil contamination may occur through the presence of naturally occurring metals and the depositing of other chemicals from human industrial activities. Various kinds of pollution also increase the concentrations of naturally occurring pollutants.

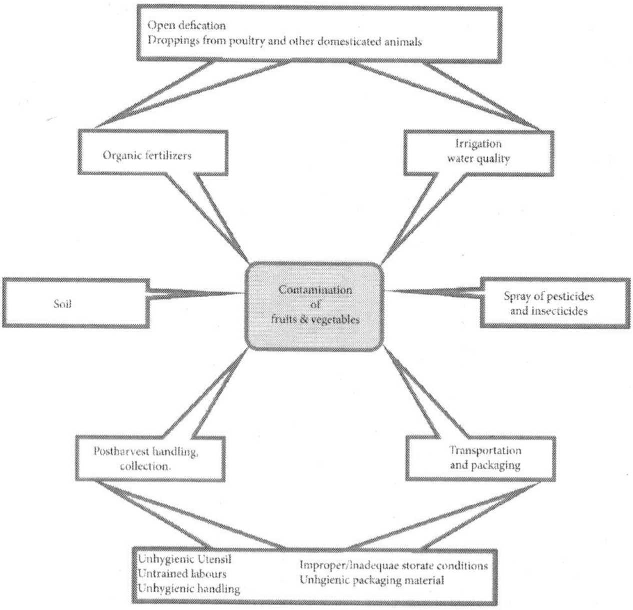

Figure 6.7: Vegetable contamination can occur through the supply of contaminated water into soil.

Source: Gadafi Iddrisu Balali, Denis Dekugmen Yar, Vera Gobe Afua Dela, Priscilla Adjei-Kusi, "Microbial Contamination, an Increasing Threat to the Consumption of Fresh Fruits and Vegetables in Today's World", International Journal of Microbiology, vol. 2020, Article ID 3029295, 13 pages, 2020. https://doi.org/10.1155/2020/3029295

The mechanism of exposure to soil contaminants through the ingestion of contaminated vegetables is as follows: soil or water entering farmlands is contaminated. The contaminants find their way into the food chain through absorption. The vegetables are harvested and sold to the markets after which they are prepared and ingested. The grievous health risks of exposure to contaminants have led to various governments placing laws that will cause a reduction in exposure to such chemicals. A good example is the Taiwan government which in 1981 began its activities to control soil quality in agriculture. From its studies, it was noted that about 1024 ha of agricultural lands were contaminated. Of the 1024 ha of land, about 319 ha was noted to have heavy metals. Over the years, there has been an increase in concern over the presence of heavy metals in soil. This is because they tend to remain in soil for longer durations. They are also not easily metabolized after they are ingested. This makes it easy for these metals to accumulate in various metabolic products such as the kidney and liver. This leads to the individual displaying various toxic symptoms. When an individual consumes vegetables with large amounts of cadmium may suffer from severe irritation in the stomach accompanied by symptoms such as diarrhea and vomiting (Salesa &Cerdà, 2020).

When individuals continue consuming vegetables with contaminants for longer durations may cause diseases such as kidney disease as the renal tubes are damaged impeding the absorption of minerals such as calcium. This makes the individual's bones very brittle and are less resistant to breakage. Some of the metals found in soil are useful in plants as essential parts of the body. When the body is exposed to high doses of these metals it may result in various health problems which include fetal retarded growth, vomiting, irritation of the throat and respiratory tract and increased risk of Alzheimer's disease (Hu et al., 2017). Plants absorb various useful nutrients, water and metals through their pores and roots, they also absorb very heavy metals through those channels.

Studies show that increased levels of soil contamination correlate to the uptake of zinc, manganese, copper and cadmium. One of the ways through which contaminants find their way into soil is through irrigation water. Studies also show that experiments on agricultural land located close to contaminated sites or through the application of untreated sewage on crops had an impact on lands and crops a sign of both man-made and natural pollution. In China, it was noted that there are relatively high accumulations of metals in vegetables and rice in agricultural sites located in close proximity to mining areas. Consumption of vegetables is very important as they are useful sources of several essential micronutrients such as vitamins.

These vitamins have the potential to regulate the immune system and inhibit oxidative stress useful in preventing chronic diseases such as cancer and cardiovascular disease. However, exposures of plants to metal elements have caused a deviation in levels of plant antioxidants. Various field and pot experiments were conducted to evaluate the link between plant growth and heavy metal pollution.

In experiments where soil containing cadmium through artificial contamination demonstrated the bioaccumulation factor. In plants such as cabbage, it was noted that there are increased levels of cadmium in soil. This led to increased concerns about heavy metal exposure to plants in the food chain. Experiments involving two planted vegetation show that contaminated soil samples extracted from an abandoned mine in Hungary showed that there are accumulated element concentrations in the vegetables (Rodrigo-Comino, 2018). When examined, the vegetable products were said to have negative effects on the health of consumers. In most cases, pot experiments are used more often when compared to field experiments as they are more effective, cheaper and easier. They are useful in indicating whether land remediation can be useful in reducing levels of heavy metals that may be needed in plants in small amounts. Accumulation of large amounts of metals in vegetables may affect antioxidant levels in the same plants. By assessing the potential risks of exposing humans to contaminated vegetables, the relevant stakeholders are able to develop management strategies needed when farming on contaminated land.

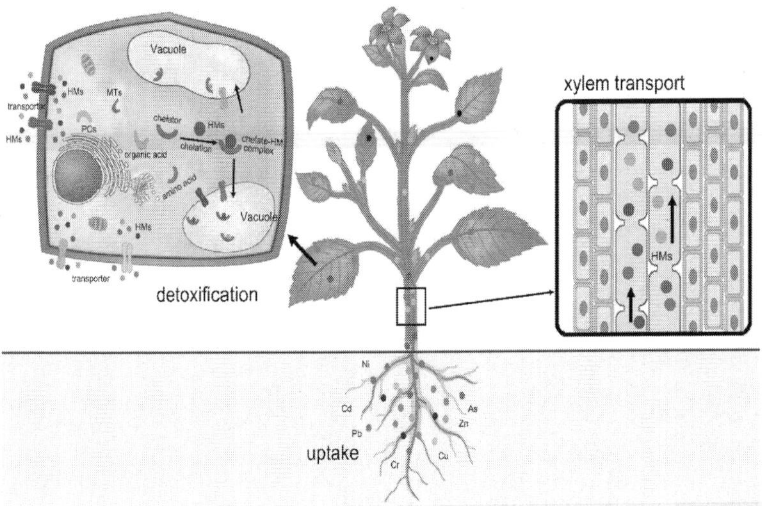

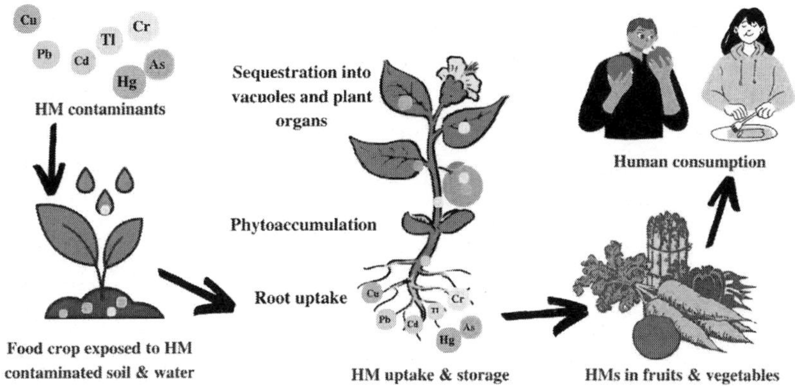

Figure 6.8: Presence of large amounts of metals in foods is linked to the presence of contaminants.

Source: C.R., V., Sankhla, M. S. , Singh, P., Jadhav, E. B. , Verma, R. K. , Awasthi, K. K. , Awasthi, G., & Nagar, V. (2022). Heavy Metal Contamination of Food Crops: Transportation via Food Chain, Human Consumption, Toxicity and Management Strategies. In H. M. Saleh, & A. I. Hassan (Eds.), Environmental Impact and Remediation of Heavy Metals. IntechOpen. https://doi.org/10.5772/intechopen.101938

One of the factors taken into consideration when farming on contaminated land is the bioaccumulation factor. The bioaccumulation factor can be defined as the ratio of the heavy metal concentration in the edible parts of vegetables to the corresponding soil samples. The bioaccumulation factors are useful in confirming the translocation capability of various heavy metals from soils to vegetables. The Bioaccumulation factor in heavy metals can be calculated by dividing Cp and Cs (Cp/Cs). Cp refers to heavy metal concentration in vegetables while Cs refers to heavy metal accumulation. Various health risk assessments have been done for various vegetables. Studies (Bifeng et al., 2017) were aimed at weighing the effects among various consumers of different ages. Comparisons were made between metal concentrations in cultivated plants and food intake data. Data from the Taiwan National consumption Database where the average daily dose were used in measuring the amounts of daily metal intake associated with local inhabitants' vegetable consumption. The average daily dose can be calculated by multiplying the concentration of metal in the vegetable by the absorption rate and also the daily intake of vegetables (Rodrigo-Comino et al., 2018).

This value is very useful in the directtt quantification of the variability and uncertainty for each parameter probability function with a given iteration size. The sum of the HQs exposed to the same pollutant by consuming different foods gives the total non-carcinogenic risk hazard index. If the risk hazard index is less than one then the non-carcinogenic risk of the substance has low significant impacts on human health. Various vegetables are used in studies and water spinach was found to have high concentrations of contaminants. It was also noted that accumulations of cadmium in soil were at higher levels compared to copper accumulation. The presence of cadmium in contaminated soils had a significant effect on the yield and growth of sweet potatoes and potatoes. Among sweet potatoes, significant effects on their growth and yields were affected by copper contamination. When sweet potato leaves and water spinach were planted in soil contaminated by heavy metals demonstrated high concentrations of vitamins E and C. this was the opposite in tomatoes and sweet potatoes. Among sweet potatoes and spinach, non-carcinogenic HI 0f Cd exposure was found to be greater than 1 while exposure to copper was less than one.

According to Hu et al., (2017), it was noted that consumption of soil contaminants through ingestion of vegetables grown in contaminated soils caused greater risks among children as they are the most vulnerable to severe effects of heavy ingestion of contaminants and metals. Low contaminant rates of contaminants in sweet potato leaves, tomatoes and potatoes showed their sustainability for planting in contaminated soil including soils found close to metal industries in replacement of other leafy vegetables with a tendency of high accumulated character for metal contaminants. Water spinach takes up large amounts of contaminants. Exposure to such kind of plants makes human beings very vulnerable to ingestion of contaminants.

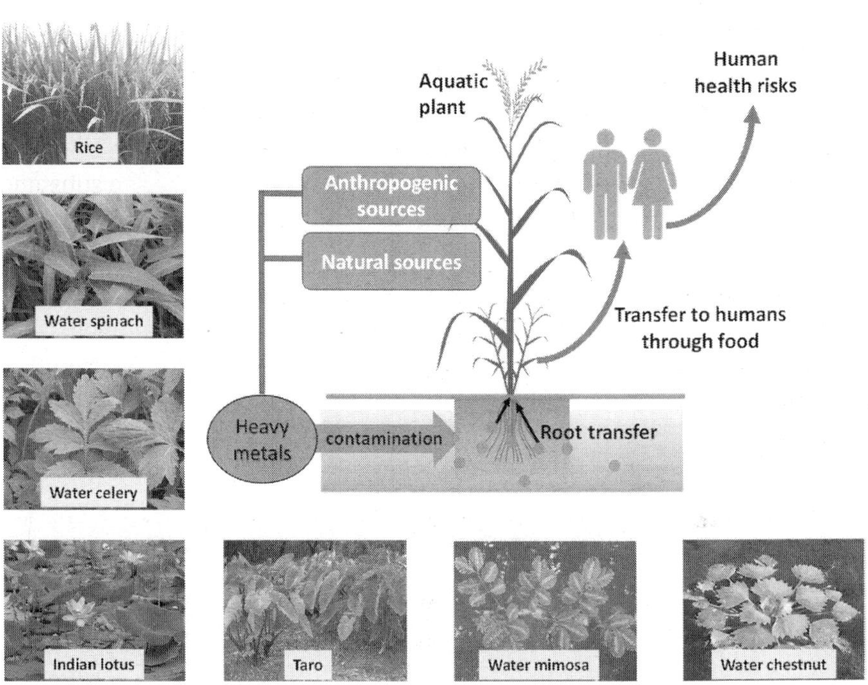

Figure 6.9: Some vegetables in the marketcontain contaminants from the soil.

Source: Uddin MM, Zakeel MCM, Zavahir JS, Marikar FMMT, Jahan I. Heavy etal Accumulation in Rice and Aquatic Plants Used as Human Food: A General Review. Toxics. 2021; 9(12):360. https://doi.org/10.3390/toxics9120360

Other than ingesting vegetables grown in contaminated soil, human beings can also be exposed to contaminants by getting in contact with contaminants. Contamination can occur through various processes before the food gets to the mouth of the user. Contamination can occur during production during the harvesting process. Vegetable products can come into contact with contaminated soil during the harvesting process mostly through the settling of dust products on plant surfaces. For instance, chicken may consume contaminated soil which will find its way into their body thereby getting into eggs. Their meats will also get contaminated. Spraying of fields with various chemicals can also make food substances to be contaminated. Fish living in soil-contaminated ponds may feed on contaminated soil particles paving way for the contaminants into foods. When such fish products are eaten, human beings are exposed to the contaminants (Rickson, 2014).

Vegetables and food products can also come into contact with contaminants through processing. Processing is the changing of plant or animal stuff into products that can be purchased as food. Processing involves various kinds of steps which include washing, sorting, trimming, slicing or shredding. Milk substances are processed through pasteurization. In most cases, those involved in the pasteurization process are required to maintain high levels of hygiene. However, if they step on contaminated soil and enter pasteurization buildings then the milk products will be contaminated. Nuts are processed through roasting, chopping or grinding. All of these processes if not done carefully can pave way for soil contaminants to find their way into food.

Processing of animals involves slaughtering them. If all the individuals involved in these processes do not properly clean their boots or various equipment then the meat will be contaminated. Wind could also blow dust into meat processing areas and exposed meat could be contaminated. Dust on any of their clothing or equipment could cause the products to be contaminated. Contamination could also happen during smoking or drying of meat products such as fish. In local areas, fish are preserved by sun drying. When the fish products are not properly covered, they can be contaminated by dust particles. Vegetables and food items can also be contaminated during the distribution process. In most cases, vegetables or fruits may not be covered when they are transported to the local markets. This leaves them open for contamination during transportation (Poesen, 2018).

There are cases where transportation may involve making one trip to the consumer or the local market while in others, there are several trips made. In the case where several trips have to be made then there is a high likelihood of the products being contaminated. During the transportation process, contamination is mostly through the spread of contaminated dust. Contamination can also occur when contaminated soil particles from the handler's boost get to food samples or when trucks used in transporting such produce are not properly cleaned.

Contamination can also occur during the preparation of various food items. In this case, contamination may occur in kitchens of homes and restaurants. Vegetables and other food products can be exposed to soil contaminants through dust or when there are contaminated soil particles on food substances and they are not properly cleaned. Ingredients exposed to soil contaminants when used in preparing dishes may cause contamination.

6.3 SKIN CONTACT WITH DUST OR SOIL

Skin contact with soil contaminants is another exposure route is often known as dermal exposure. This kind of exposure happens when skin comes into direct contact with contaminated soil or dust. Dermal exposure can occur in various ways including water, sediment, outdoor soil or dust and indoor dust that has settled on various objects found. With regards to water. Dermal exposure happens when water used in various activities such as swimming, washing and bathing has come into contact with contaminated soil particles. This is a common occurrence for individuals living near rivers or lakes located on contaminated soils. As they swim in such waters, their skin comes into contact with soil particles in water. Dermal exposure can also occur through exposure to outdoor dust as individuals take part in recreational activities more so in windy areas. This kind of exposure occurs mostly with children as they are prone to conduct various activities outdoors. As they play in soil, their skin comes into contact with it. If the soil is contaminated, then the contaminants will find their way into their bodies.

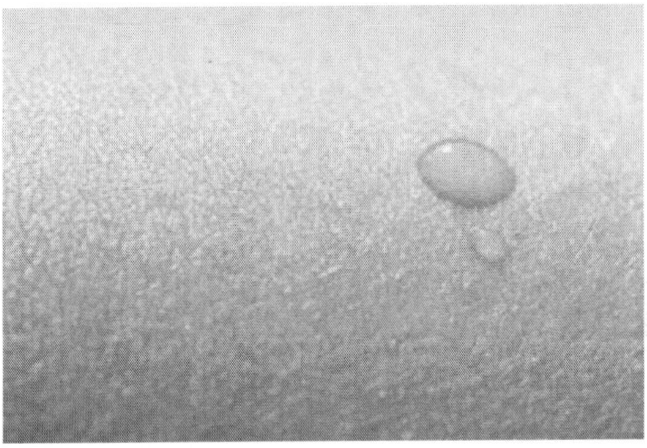

Figure 6.10: Dermal exposure is exposure to soil contaminants through the skin.

Source: https://www.livescience.com/27115-skin-facts-diseases-conditions.html

There is a higher risk of exposure to contaminants during recreational activities on beaches. Beaches whose soil is contaminated will increase the risk of exposure to larger numbers of people as several visit beaches for recreational activities. As children play with the soil or burry each other in soil, their skin is exposed to contaminants. Activities like gardening, also provide a pathway for exposure to soil contaminants. During such activities, the gardener may touch their face or other parts of their body allowing skin contact with the contaminants. Activities such as construction and mining involve workers getting in contact with soil among other equipment. If the soil they are using during these activities is contaminates then their skin is at risk of exposure is they do not wear the proper attire. In some cases, the protective clothing may be compromised allowing soil and other particles into the body. They are also exposed to dust due to the displacement of soil. The dust particles may settle on their skin more so on hair enabling prolonged exposure to contaminants (Pijl et al., 2020).

While indoors, human beings can also be exposed to contaminated soil mostly through dust. As wind blows, contaminated dust articles find their way into objects and surfaces in their homes. Some of these particles settle on carpets or floors hence dermal entry through the legs. Some dust particles settle on clothing while others settle on counter tops and other surfaces. As the individual living in the home touches different surfaces, the dust particles settle on their skin. This could also happen as they are cleaning the surfaces. There are various extents of exposure due to dermal entry. Some factors affecting the extent of exposure include the concentration of the contaminants in the medium. When very concentrated soil particles land on skin, high levels of contaminant exposure are anticipated. Other factors include skin surface area exposed, film thickness of solids and small particles on skin, dermal adherence to solids on skin and residue transfer factors.

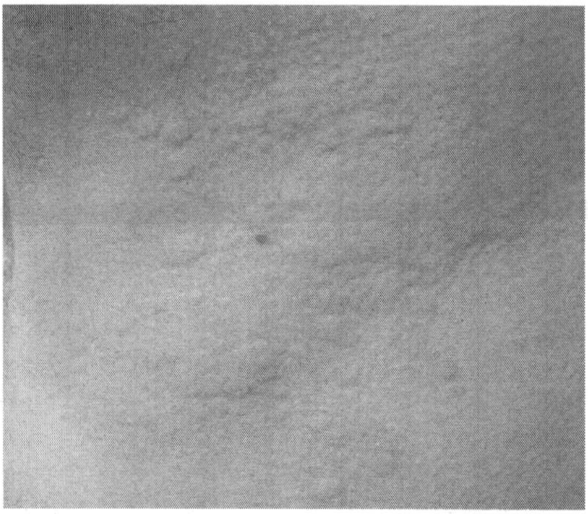

Figure 6.11: Dust particles settled on skin surfaces may be absorbed into the body.

Source: https://en.wikipedia.org/wiki/File:Hives2010.JPG

The amount of contaminants exposed to the body can be calculated through various methods. Assessments on dermal exposure are done in a two-step process. The first step involves contact between skin and the contaminant. The second step involves the evaluation of receptor absorption into the body through skin. The amount of contaminants available for interaction with targeted tissues and organs is dependent on the general amount of contaminants present. Media-specific contaminant concentration is influenced by the magnitude of exposure and also the time frame of exposure. It is also affected by the surface area available for contamination (Pham et al., 2018). Some of the terms used in characterizing dermal exposure to soil contaminants include potential dose which is the number of contaminants exposed to skin some of which are absorbed into the body. The applied dose is the term used to refer to the amount of contaminants present at the absorption barrier and absorbed into the skin. Internal dose is a term that refers to the number of contaminants that have already been absorbed into the body and are available for interaction with tissues and organs.

With dermal exposure, the potential dermal dose covers the contaminant and the matrix in which it is suspended (soil) as it comes into contact with the skin. Studies show that it is highly unlikely for the total amount of potential dose to come in contact with skin. For instance, when wind blows and carries away contaminated soil it is highly unlikely that all the soil carried away will come in contact with skin. The amount of contaminants in soil that comes in contact with skin is referred to as the applied dose. The amount of contaminants in soil that finds their way into the body after penetrating the skin is known as the internal dose. The internal dose has the ability to enter the bloodstream and interact with tissues and organs causing various biological effects. The rate at which contaminants are absorbed into the body determines the absorbed or internal dermal dose.

The amount of contaminant that penetrates skin per unit time is known as flux. It is well explained under Fick's law of diffusion for membranes. The law states that the flux of contaminants across a homogenous membrane is proportional to the concentration difference between the membrane's inner surface and the membrane outer surface and inversely proportional to the thickness of the membrane. This law makes a number of assumptions including the skin being a homogenous membrane, the flux of contaminants across skin is primarily due to passive diffusion and the permeability coefficient is specific to the vehicle contacting the skin. For dermal contact with soil, the concentration of contaminants coming in contact with skin is usually a function of the concentration of the contaminants in soil and the amount of solid that adheres to skin per unit of surface area.

When estimating the concentration of contaminants on skin, the concentration of contaminants in soil may be multiplied by the adherence rates. Absorption rates are usually taken into consideration when estimating the amount of contaminants in the body. Levels of contaminants in the body are influenced by the exposure frequency which is the number of times skin is exposed to soil contaminants. The frequency could be daily, weekly, monthly or annually. Exposure duration is also another factor. It refers to the amount of time an individual is exposed to contaminants being evaluated. The average time is the amount of time over which exposure is averaged (Beilicci, Beilicci &Stefănescu, 2014).

Once the body is exposed to soil contaminants such as dust or as soil, they move into skin through a combination of processes namely contact transfer and diffusion. Under contact transfer, the spread of contaminants is very fast as the contaminants are directly transferred from particles directly touching the skin on to the skin. Under diffusion, contaminants in both water and air-filled pores occur more slowly. However, it accounts for a larger percentage of the total absorption. There are various tools used in differentiating the two processes. There are various experiments done to evaluate how the two processes influence blood contamination.

Particle size is a huge determinant of the rate of exposure to contaminants. Small-sized soil particles are able to easily penetrate skin. The opposite is noted in large-sized particles. The largest particle size that can penetrate through skin is 150 micrometers. The manner in which contaminated soil is distributed in the body also influences the entry of contaminants into the bloodstream. Uniform distribution of soil contaminants on the skin may influence levels of contaminants in the bloodstream. Saturation of chemicals in soil is usually taken into consideration. Longer durations of exposure to contaminated soil increase the risks of contaminant effects.

6.4 VAPORS FROM THE SOIL

Exposure to soil contaminants through vapors from contaminated soil is known as soil vapor intrusion. This process involves volatile chemicals moving from soil sub-surfaces into the indoor air of overlying buildings which are later on inhaled by human beings. Vapor from soil is also known as soil gas and is at times defined as the air found between soil particles. Vapors from soil can easily be inhaled by human beings as they go on with their daily activities. Some of the vapors enter buildings where they work or reside through cracks in slabs or basement floors. They can also use openings through sump pumps or electrical wiring. The vapor is further drawn into buildings when a negative pressure is created through air-conditioning systems, ventilation and heating. Entry of soil vapor may be illustrated by the seeping of radon gas into buildings. This kind of exposure to chemicals occurs for a long duration of time without noticing and can affect larger populations of people.

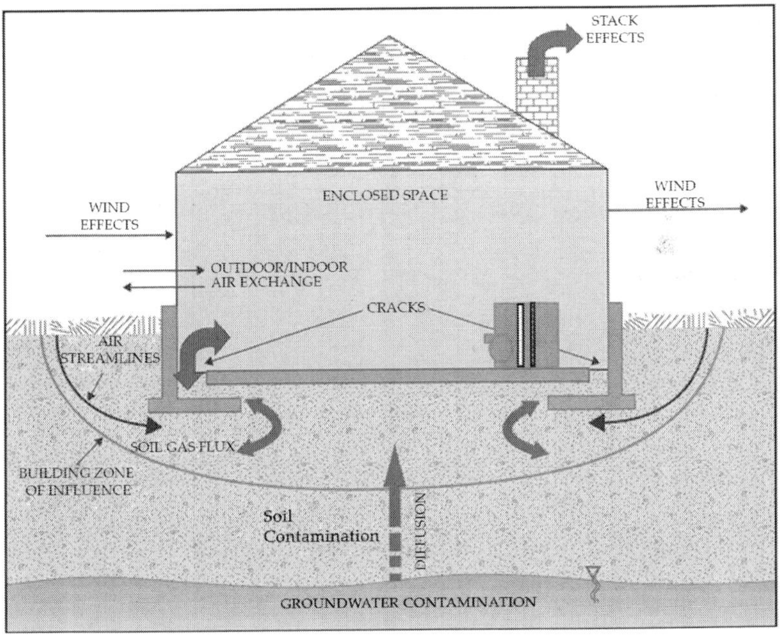

Figure 6.12: Vapors from soil are mostly released through evaporation.

Source: https://www.enviro.wiki/index.php?title=File:Lutes-Article1w2.Fig_1.PNG

The release of vapors from soil can occur without any individual noticing as environmental processes such as evaporation usually go unnoticed. When contaminants from various sources seep into soil, they may settle into soil moisture contents, soil air or the solid part of soil. Heating occurs when soil is exposed to large amounts of sun, which causes the soil to expand causing the release of contaminated soil air. Vapors can also be released from the evaporation of soil water. Soil vapor rises and mixes with other air in the atmosphere and air currents cause its movement over large areas.

This kind of exposure to soil contaminants is mostly through inhalation as the contaminate vapor enters the body through the nose (Zhao et al., 2015).

As mentioned, soil vapor may be formed through the evaporation of soil water filled with contaminants. Significant amounts of the contaminants could also exist in their natural form. If they easily evaporate as soil vapor then the contaminants are volatile in nature. A good example of volatile soil contaminants is volatile organic compounds. There are various sources of volatile soil contaminants including contaminated ground water and buried wastes. Contaminated soil vapors may exist in the ozone layer (Pietro & Mahajan, 2022). This is because some small amounts of contaminated soil vapor find their way into the upper atmosphere. It interacts with other gases in the upper atmosphere which will increase or decrease the concentration of the contaminants. Their presence in the atmosphere means that people are at a continuous risk of exposure to contaminated soil vapors.

This kind of exposure can reach a large population of people as it has the ability to spread from sources of the contaminated soil vapors into areas where there is no kind of contamination. Exposure to this kind of contaminants is highly likely in that the vapors can settle on walls of buildings among various surfaces in the environment increases the likelihood of interaction with these kinds of vapors. However, depending on the concentration of the contaminants and the type of contaminants present in air, health effects will vary. Exposure to high concentrations of contaminants over longer durations of time increases the likelihood of an individual suffering from any health effects. If the contaminants are very lethal the health effects may be noted among individuals living in that area.

With soil vapor, various kinds of chemicals could be in the atmosphere among them volatile organic compounds, chemicals from industries and solvents for various processes. Some of the contaminants are at times petroleum related and include styrene, xylenes, ethyl benzene and benzene. It is important to note that contaminated soil vapors may not be the only source of air contaminants is the atmosphere. Soil vapors account for a small portion of contaminated air in the atmosphere as they are greatly dependent on the amount of heat and other factors that will allow the escape of vapor into the atmosphere. When compared to other forms of exposure to soil contaminants, soil vapors are very dangerous. They could exist in the atmosphere for relatively long durations without anyone noticing hence there is a large likelihood of continuous exposure to the contaminants. They can also do more damage to larger groups of people when compared to other forms of exposure.

In most cases, areas that have recorded the release of soil vapors are those whose soils and groundwater have been noted to have significant amounts of contaminants. In most cases, the sites are identified by the state or government and clean up measures are put in place to minimize risks of contamination. In most cases, there are different methods employed by different states in investigating the presence of soil vapor in the environment. The most common method involves the use obtaining of several air samples (Priori et al., 2020). The samples are extensively analyzed to determine the extent of vapor contamination. There are four main samples obtained when analyzing the presence of contaminated soil vapor in the atmosphere. They include soil vapors samples, sub-slab vapor samples, outdoor air samples and indoor air samples. In the case of soil vapor samples, these samples are collected when the analysis is done to characterize the nature and extent of vapor contamination in a given region. These samples are usually collected before indoor air samples or sub-slab vapor air samples. These kinds of air samples are used in determining the potential of human beings being exposed to contaminants. It is important to note that soil vapor samples are not the same as soil samples.

Sub-slab vapor samples are collected when there is a need for an analysis to characterize the nature and extent of vapor contamination in soil found beneath a slab or basement foundations. The results from these samples are often used in determining the current and future human exposures. From the study, it could be established that future exposures could occur when a crack develops in the building's foundation or the installation of an air-conditioning system which will make the movement of contaminated soil vapor within the building. Indoor air samples are collected by scientists when characterizing the nature and extent of air contamination in a building. These samples are useful in evaluating whether there are current human exposures. After collection, these samples are compared to outdoor air and sub-slab vapor and are useful in determining the source of the volatile chemicals. The final category is the outdoor air samples. These kinds of samples are usually collected when characterizing site-specific background air conditions. The results of such samples are used in evaluating the extent to which outdoor air sources may be contributing to the entry of contaminated soil vapor into buildings.

If the results of the samples show that contaminated soil vapors are found in buildings. Further investigation may be done and further monitoring will be done. Mitigation actions are also recommended. Further evaluation is done so as to determine the extent of soil vapor contamination (Pimentel

& Burgess, 2013). Monitoring is usually done when there is a potential for soil vapor intrusion to occur in buildings or when there are changes in the environmental conditions. Some mitigation actions involve sealing cracks in buildings and covering contaminated soil with layers of uncontaminated soil to avoid the release of soil vapor.

6.5 INHALING DUSTS WHILE WORKING IN SOIL OR WINDY ENVIRONMENTS

Among the most common routes of exposure to soil contaminants is through inhaling contaminated dust in their areas of work. Dust in this case is comprised of soil particles whose diameter is less than 75 micrometers. The suspension and settling of dust is determined by the weight of the dust particles. In the environment, contaminants find their way into soil. They fuse with soil rendering soil contaminated. In windy areas, wind carries away soil particles from areas of contamination. As the wind blows, some of the dust particles in inhaled by individuals and enter their respiratory tract. For dust to be airborne, the most important factor is particle size. The behavior of contaminated soil particles while in the air is determined by the particle's diameter. The diameter of soil particles affects the movement of contaminated soil particles through the lungs and their depositing at different sites of the respiratory tract.

Figure 6.13: Exposure to soil contaminants in workplaces is known as an occupational hazard.

Source: Paula, Ricardo & Ayres da Silva, Lineu & Ayres da Silva, Anna. (2016). haracterization of the unhealthy and hazardous situations under the forensic expert survey and safety management in quarries:. REM - International Engineering Journal. 69. 361-366. 10.1590/0370-44672015690011.

In the workplace, there are different kinds of dust inhaled by workers. The presence of contaminants in the dust particles makes the health risks more adverse. There can therefore be four main groups of dust contaminants in the workplace. They include mineral dust. This is mostly inhaled by individuals working in quarries and construction sites. Contaminated dusts in these areas are mostly made of cement, coal or crystalline silica. There is metallic dust inhaled by individuals working in quarries or construction sites. Dust inhaled in these areas includes beryllium, nickel, cadmium and lead dusts. There are chemical dusts including pesticides and bulk chemicals, organic dusts including pollen, dust, tea, cotton, wood and flour and finally biohazards including spores, molds and viable particles. In the workplace, exposure to dust is covered under occupational hygiene. Under occupational hygiene, workers are expected to wear protective clothing such as masks to protect them from dust (Nearing, Lane, & Lopes, 2017).

As they go about their mining activities, they cause disturbances in the soil causing dust to risk in the tunnels. If they had not worn any masks, there is a high probability that they will inhale large amounts of contaminated dust. The same applies in the construction industry. As they dig such contaminated soil, soil particles rise and can be inhaled. Inhalation of contaminated soil particles can occur through the nose hence the nasal route or through the mouth hence the name oral route. Inhaling of contaminated dust is dependent on particle aerodynamic diameter, breathing rate and movement of air around the body. Once in the lungs, the inhaled contaminated dust can either be deposited or exhaled as influenced by particle-related or physiological particles.

When being deposited, the five main mechanisms of deposition include electrostatic deposition, interception, diffusion, inertial impact and sedimentation. With inhaled airborne dust, the two main important mechanisms allowing entry of dust into the lungs are impaction and sedimentation. Both processes are governed by particle aerodynamic diameter. With regards to the aerodynamic diameter, the largest diameters of particles inhalable are about 30 micrometers. These kinds of particles are deposited in head airways and entry routes of the larynx. Dust particles are deposited in the nose through filtration by nasal hairs. Impaction in the nose causes changes in the direction of airflow. Individuals are advised to breathe through the nose rather than the mouth as in the nose, there is better filtration. For those who breathe through the mouth, large amounts of the contaminants will be deposited in the lungs (Nigussie et al., 2015).

When particles are not deposited in the nose, they are deposited in the tracheobronchial airway region which will later on find its way out through mucociliary clearance. Very small particles further penetrate into the respiratory system through dissolution. The very fine dust particles penetrate further into the alveolar regions. This is usually the region where inhaled gases are absorbed into the blood. This is the furthest region of dust penetration. When dust particles are less than 2 micrometers, then larger amounts of dust are deposited in the alveolar region.

Large amounts of inhaled dust particles are exhaled without being deposited. However, the little amounts of deposited dust particles are sufficient to cause health complications that are respiratory related. Once inhaled, some of the contaminants could eventually find their way into the bloodstream and finally into various organs and tissues.

6.6 EXPOSURE TO CONTAMINATED GROUND WATER

Exposure to soil contaminants through contaminated groundwater is mostly an oral kind of ingestion. It is very easy for groundwater to be exposed to various contaminants. As it flows, groundwater can interact with contaminated soils. Some of the soil particles can remain in water as it is pumped to various facilities. In most cases, after collecting groundwater, it goes through numerous processes seeking to remove soil particles and other contaminants found in water making water safe for drinking. However, there are cases where the filtration process is not properly done and some of the soil particles in the water contaminate the entire water supply. With groundwater, the contaminants may not be highly concentrated as water will dilute some of the contaminants. However, there are communities and groups of people living in areas that are not advanced. This means that their water supply does not undergo any filtration process. Therefore when contaminated groundwater finds its way to rivers or lakes, the entire water body has soil contaminants (Molla& Sisheber, 2017). Water fetched from these lakes or rivers is used in performing various duties including cooking, washing and drinking. As it is not filtered to remove soil, there is a higher likelihood that those individuals could consume contaminated water. Such water is used in cleaning utensils used in eating meaning that contaminated soil particles could be present on the utensils or as they are used in cleaning foodstuffs, soil particles could be deposited in them and ingested. These contaminants enter their digestive system and finally find their way into the bloodstream. They are then deposited into various tissues and organs. Therefore the four main ways of exposure to soil contaminants in through inhalation, ingestion and dermal exposure.

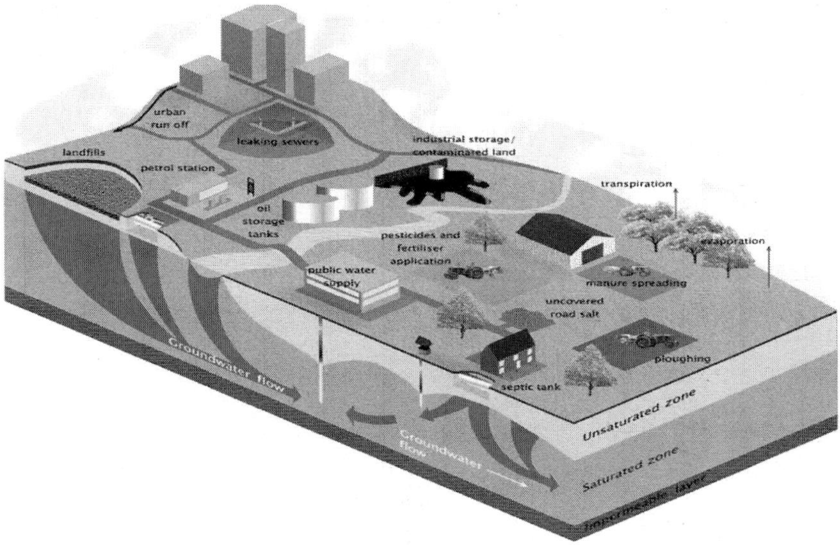

Figure 6.14: Groundwater contaminated with soil particles is hazardous to human health.

Source: Wang, H., Liu, S., & Du, S. (2013). The Investigation and Assessment on Groundwater Organic Pollution. In (Ed.), Organic Pollutants - Monitoring, Risk and Treatment. IntechOpen. https://doi.org/10.5772/53549

CHAPTER 7

SOIL CONTAMINATION AND SITES OF SPECIAL CONCERN

CONTENTS

7.1 Industrial And Manufacturing Sites 168

7.2 Industrial Activities 170

7.3 Manufacturing and Recycling 172

7.4 Transport 173

7.5 Workplace Accidents 174

7.6 Landfills, Junkyards and Waste Disposal Sites 176

7.7 Types of Landfills 178

7.8 A Specially Designed Landfill 180

7.9 Highway Corridors, Parking Lots, Areas of Heavy Traffic 180

7.10 Pollution 181

7.11 The Most Effective Gardening Method 190

7.12 Former Farmland with the Build-Up of Contaminants 191

7.1 INDUSTRIAL AND MANUFACTURING SITES

Pollutants are frequently found in industrial and manufacturing facility soils. Many types of contamination may occur depending on the factory's output. Chemical leaks from buildings or vehicles can be harmful to the environment. The facility may contain a polluted waste storage section or holding area previously thought to be safe. In addition, industrial sites can be pretty significant. Full-site soil remediation is an unavoidable, costly, and time-consuming process.

As it creates products for industrial use from raw materials, industrial manufacturing output supports subsequent mass production in the majority of other sectors. It is in charge of a wide range of industrial products, including hardware, paper and packaging, glass, and various metal and plastic types (Moges &Holden, 2007). Regardless of the products available, the goal is always the same: to reduce or eliminate the time and effort required by humans to complete a task. A large amount of equipment is required in every economy to produce many goods and services in a timely and cost-effective manner.

Figure 7.1: Industrial manufacturing facility plant modern factory.

Source: By Photo: Andreas Praefcke - Self-published work by AndreasPraefcke, CC BY 3.0, https://commons.wikimedia.org/w/index.php?curid=111097

Industrial equipment can be classified as standard or custom-built based on its specifications. Standard equipment is less expensive to manufacture and may be utilized in various sectors. The issue with this approach is that it does not always meet the specific needs of newly established businesses. As a result, specialized equipment is more expensive and more profitable. It takes longer to make, but it can be tailored to meet the needs of individual customers. Industrial equipment is classified into seven categories. Special-purpose machinery includes agricultural, construction, mining, industrial, commercial, and service machinery (Mekuria et al., 2009). The remaining four categories include various equipment used in various industries, such as ventilation, heating, cooling, metalworking, engine, and engine-related equipment, and miscellaneous general purpose gear. Many of these companies are multi-product conglomerates that serve many markets. These conglomerates are typically the result of prior company mergers and acquisitions.

Mining and industrialization have historically played a significant role in polluting the land. Trace elements and organic pollutants are more prevalent in industrialized areas than in non-industrialized areas. Industrial operations contribute to a wide range of environmental issues due to both planned and unintentional pollution of the environment.

Several ethical industrial enterprises go to great lengths to reduce the possibility that their actions will harm the environment or public health. They are not alone. Several countries worldwide have legislation restricting industrial activity and ensuring pollution control. Several legal systems around the world support the idea of using best practices in mining, manufacturing, recycling, and disposal. "Best practices" numerous international organizations and industry associations are working to develop standards and guidelines to assist various sectors in improving their safety and sustainability while reducing their negative impact on the natural environment and human health. UNEP, UNIDO, and the Chemical Conventions are a few examples (Meshesha et al., 2012). No matter how hard we try, industrial activity continues to degrade the soil, particularly in low and middle-income countries. Workplace injuries and illnesses are also a significant source of pollution. Another problem is the lack of environmental management of previously abandoned industrial sites and contamination from those sites and trash disposal sites. All of these continue to pollute the environment.

Chemicals are an excellent example of how global industrial output is likely to rise in the future. Between 2000 and 2017, it is estimated that the world produced 2.3 billion tonnes of chemicals per year. Petroleum, special chemicals, and polymers made up 26 percent, 27 percent, and 22 percent, respectively. Aside from medicines, chemical consumption is expected to increase by 85 percent by 2030, with China and the EU continuing to be the primary users. According to estimates, each person on the planet would have spent approximately USD 500 on chemicals in 2011 (Ma, Zhao, & Zhu, 2021). They have harmed the development of human civilization as a result of unintentional human and environmental exposure and toxicity to synthetic chemicals.

After mining and quarrying, the following most significant sources of industrial soil contamination are manufacturing, energy production, facility construction, and transportation.

7.2 INDUSTRIAL ACTIVITIES

Humans have been mining since the beginning of time. Smelting and waste rock deposits, as well as tailings and other byproducts and emissions generated during the processing of mined minerals, are significant sources of trace elements in addition to the mining process itself. Mining and mineral processing companies that are responsible strive to reduce the environmental and public health risks associated with their operations. The International Council of Mining and Metals (ICMM) is one of several organizations working to improve the mining sector's safety and sustainability while minimizing its environmental impact. Despite this, the mining and metal processing sectors continue to harm the environment, particularly in low- and middle-income countries (LMICs) (Lim et al., 2020).

Many rocks contain trace elements that are remobilized during mineral deposits. Mining wastes still contain trace elements but at low levels to make extraction cost-effective while posing a risk to the environment and human health. Chemical separation of the target resource may be required during precious metal extraction, resulting in large volumes of chemically enriched waste (particularly cyanide solutions).

Figure 7.2: Mining, Quarrying, and Oil and Gas Extraction.

Source: https://www.namesnack.com/guides/business-names/mining-quarrying-and-oil-and-gas-extraction.

Crushing, powdering, and other processes that result in mined ore tailings liquid slurries pose a significant problem. Mineral particles are present in these liquid slurries. On-site, a dam is commonly used to build settling ponds where waste is collected. After the sediments have dried, the water is either recycled or released. The ore enrichment process produces tailings that contain unwanted and leftover trace elements. Soil contamination is primarily caused by tailings dam failure or wind erosion, which disperses small mineral particles from the surface if vegetation or a capping layer is not entirely covered. As a result, pollution may spread to areas not previously affected. According to Rico, there were 147 tailing dam ruptures worldwide in 2008 (Li et al., 2018).

The majority of these incidents were the result of poor maintenance. Accidents have occurred in Spain, Romania, and Brazil, to name a few. Hazardous materials containing arsenic, cadmium, copper, and zinc were discharged into soils and surface waterways due to these incidents. Leaching from mining and rock waste dumps and wind erosion can introduce trace elements into the soil. Acid mine drainage (AMD) can form when reduced sulfur minerals in mining waste, such as pyrites, are exposed to oxygen and water. The environment's acidity increases metal solubility in other minerals containing sulfides, gangue silicates, and carbonate components. Massive amounts of very acidic wastewater laced with high levels of sulfates, iron, and trace metals are released into the environment in the event of a leak.

Aside from spillage during ore concentrate transportation, unintentional overspill is another potential source of soil contamination. Trace element contamination in soils can occur locally and globally if smelter emissions of particulate and gaseous pollutants are not adequately abated. Trace elements may be present in ore concentrations supplied to smelters if they come from a regionally limited source.

According to the FAO and ITPS, mining coal, gold, uranium, and wolfram/tungsten causes the most severe soil degradation. Compared to all other areas tested, Barapukuria, Bangladesh; Ledo, India; Ptolemais-Amynteon, Greece; and Tibagi River, Brazil had the most soil pollution from trace elements. During the extraction of gold, various hazardous chemicals can be released. Mercury is a standard gold extraction method, and it is discharged in waste tailings, where it either volatilizes or enters the soil and groundwater. In the 1980s and early 1990s, artisanal gold mining in the Amazon basin may have released over 4000 tonnes of mercury. Organic soils contain approximately 150 thousand tonnes of mercury from human activity. In contrast, mineral soils contain up to 800 thousand tonnes, with small-scale gold mining in South America, East, and Southeast Asia accounting for 15% of that total. Large-scale gold mining releases over 2,700 metric tons of mercury into the ground each year (Blake et al., 2021).

Even after mining operations are completed, mining sites can still be a source of pollution. If proper long-term management is not provided, tailing dams and rock waste dumps are prone to weathering, water, and wind erosion, allowing toxins to be dispersed into the surrounding soils. Significant levels of lead and zinc, for example, have been discovered in the soils and mining waste of an abandoned mining area in Morocco's lead district of Alto Moulouya, linked to carbonates and occluded in readily reducible manganese and iron oxides, and can easily be transported to nearby soils by erosion.

7.3 MANUFACTURING AND RECYCLING

Contaminants in the manufacturing industry vary depending on the product and the manufacturing process. Here are some examples of pollutants, their effects on the environment and human health, and the significance of treating them.

Nitrogen oxides, sulfur dioxide, carbon dioxide, organic compounds, PAHs, and trace elements are among the pollutants that stick to airborne dust particles if cement manufacturing plants are not adequately managed. During the cement manufacturing process, cadmium, copper, and zinc traces may be released into the atmosphere. The contamination of soil with trace elements from cement plant dust emissions has been documented in the literature. Trace element concentrations exceeded US EPA regulations at the Douroud cement plant, and wind direction was where the highest concentrations of trace elements in topsoil occurred (Lal, 2019). Trace element pollution affected agricultural soils near a cement factory in Nigeria, allowing harmful elements into the food chain.

7.4 TRANSPORT

Particles, trace metals such as arsenic, cadmium, chromium (as well as copper, lead, nickel, and zinc), PAHs, and road salts are among the pollutants released by vehicle traffic. Pollutants emitted by incomplete fuel combustion, oil leakage from engines and hydraulic systems (such as brakes), brake dust, road surface leaching, corrosion of traffic control devices (such as traffic lights), and road salts applied directly to roads (as in snow removal operations) are just a few of the many sources of these pollutants. When tires wear down, for example, rubber and plastic dust are significant sources of zinc. PAHs (polycyclic aromatic hydrocarbons) are found in car exhaust and asphalt-based road construction materials (Lefèvre et al., 2021). Soil and plant ion balances are disrupted during the winter months due to de-icing salt treatment, affecting trace element mobility.

Figure 7.3: Nickel.

Source: https://en.wikipedia.org/wiki/Nickel

Transportation is predicted to account for more than 63 percent of global oil use, while alternative fuel replacement is not expected to reach 5 percent by 2021. Europe anticipates electric car sales to reach 27 to 41 percent of the market by 2030 due to its fuel economy requirements and legislation restricting particle pollution. Transportation contributes to global warming, and CO2 and NOx emissions are the primary sources. As a result, particle matter and unburned hydrocarbon emissions are produced. When oil is burned, nitrogen oxides are produced, contributing to smog and acid rain.

PAHs and trace metals such as cadmium, cobalt, chromium, copper, iron, lead, mercury, molybdenum, and zinc pollute soils and plants near train tracks. Railway sleepers and ties and creosote wood treatments that inhibit decomposition are all sources of PAHs near railway lines. Metal concentrations in surrounding soils can grow due to abrasion from the wheels, rails, and pantographs of electric locomotives. Chromium, copper, iron, lead, mercury, and zinc levels in soil were also high at cleaning bays and railway sidings (Keesstra et al., 2018).

Along railway lines, copper chrome arsenate (CCA), an arsenic-based insecticide and wood preservative, can be discovered. Pentachlorophenol, which has been classified as a persistent organic pollutant (POP) by the Stockholm Convention since 2017, was used as a wood preservative in railroad sleepers and ties. As a result of alkaline discharge from roadways, local soil pH values rise, restricting trace element mobility. Trace elements can be harmful to groundwater due to downward percolation, especially in areas with acidic sandy soil and a high groundwater table.

It is typical practice in metropolitan and peri-urban regions to produce food on soil contaminated by neighboring roadways. According to much research, the metal loads of plants decrease with increasing distance from the road. The same is valid for agricultural soils located near roadways. Poisons are more likely to infiltrate our food chain and disrupt terrestrial and aquatic ecosystems if soils along food-producing sites are contaminated.

7.5 WORKPLACE ACCIDENTS

Natural catastrophes, incorrect design and administration of industrial equipment, willful damage, armed conflicts, and unexpected mishaps are all subjects of industrial accident research. Industrial accidents can happen during raw material extraction, pipeline transportation and transfer, storage, and processing. Industrial accidents can occur regardless of a country's level of development.

Figure 7.4: Common soil contamination sources.

Source: By Bengt Nyman from Vaxholm, Sweden - originally posted to Flickr as IMG_6064-1, CC BY 2.0, https://commons.wikimedia.org/w/index.php?curid=10777224

Toxic PHCs and the formation of an impermeable barrier that prevents overly high concentrations of water and gas from flowing into and out of the soil are direct repercussions of crude oil spills on soil life and ecosystems. Plant roots suffocate in this new anaerobic environment, and bacterial activity decreases. An oil leak at a petroleum drilling site can affect soils by releasing drilling fluids, crude oil, and refined petroleum products utilized in the equipment. Between 1970 and 2000, the Nigerian federal government documented approximately 7,000 spills. Oil spills in the Niger Delta area of Nigeria, which produces oil and gas, are regular. According to a survey done between 2011 and 2016, the most prevalent causes of oil leaks (74.4 percent) were vandalism and crude oil theft, followed by corrosion, equipment failure, and human error a proportion of 25.3 percent (Kelly et al., 2020).

7.6 LANDFILLS, JUNKYARDS AND WASTE DISPOSAL SITES

Like industrial sites, landfills, junkyards, and garbage disposal facilities provide a significant risk of soil contamination. These areas include various poisons, such as lead, arsenic, and petroleum chemicals. Each of them poses a risk to human safety on its own. They have the potential to produce even more hazardous chemicals when combined. Containment and cleanup of these sites are expensive, technically complex, and logistically challenging.

Figure 7.5: Landfills.

Source: By Cezary p - Own work, CC BY-SA 4.0, https://commons.wikimedia.org/w/index.php?curid=4092940

These technical recommendations provide background information on specially built landfills that dispose of wastes with one or more hazardous characteristics. To provide a more comprehensive picture of the environmental and health concerns associated with landfills, the technical recommendations look at the landfilling of other wastes, such as domestic garbage, as specified in the Basel Convention. Along with the regulations for new landfills, the technical recommendations address the significant issue of existing or abandoned dump sites, which necessitate extensive supervision, monitoring, and routine cleanup (Kheyrodin, 2016). They describe the many types of hazardous waste that should be disposed of in landfills.

Landfilling is by far the most common form of waste disposal in most countries. It is a phrase that refers to the dumping of trash into or into the ground, and it frequently alludes to long-term storage due to the nature of the materials involved. Numerous countries have implemented the "specially engineered landfill concept," in which wastes are only sent to sites with natural, augmented, or directly provided by liners containment properties. The overall engineering is such that the wastes are only sent to sites with natural, augmented, or directly provided by liners containment properties. The overall engineering is such that the wastes are only sent to sites with natural, augmented, or directly provided by liners containment properties. The overall engineering is such that the wastes are only when all other options for reducing, mitigating, or eliminating the hazards posed by such wastes are exhausted, such dumps are used as a last resort (Kumarasinghe, 2021). The first option in the hazardous waste management hierarchy is to reduce reliance on land disposal by waste avoidance, minimization, and other technological measures. Preventing hazardous waste generation and rehabilitating contaminated sites are critical components of environmentally sound management, according to UNCED Agenda 21, Chapter 20, and both require knowledge, experienced personnel, facilities, financial resources, and technical, and scientific capabilities.

Figure 7.6: Invasive species in rubbish dumps.

Source: By Truckdriver53 - Own work, CC BY-SA 4.0, https://commons.wikimedia.org/w/index.php?curid=95231741

A proactive waste management approach should be established in this instance. An outline of such a method is provided below: Even if there is insufficient evidence to demonstrate a direct relationship between inputs and outcomes, environmental protection should be attempted when there is reason to believe that chemicals, trash, or energy released into the environment will cause harm. While avoiding garbage is unquestionably the best waste management strategy, current waste development patterns indicate that landfills will continue to be used for disposal. Because of the ubiquity of unregulated disposal sites, many governments are having difficulty creating a hazardous waste management program.

They must prioritize rapid corrective procedures and improvements in areas where these facilities are still operational and waste management plans that prioritize long-term security over short-term expediency, which is ineffective when it comes to garbage landfilling (Kavian et al., 2019). Short-term programs must begin with a comprehensive site study and environmental impact assessment and monitoring and reporting on garbage collection and any issues that arise as a result of uncontrolled disposal. Control measures, as well as remediation efforts, must be implemented at dumping sites. Improved moisture content, pH, and compaction/density controls would benefit current and future landfills that dump waste that contains a high proportion of biodegradable or bio convertible materials. These improvements will make it easier to break down chemical and biochemical waste, resulting in a faster rate of bulk stabilization. It is becoming more common to expand pre-treatment methods prior to landfilling. Solidification and chemical fixing are two helpful technologies that are gaining popularity (Jomaa et al., 2013). Immediately following (or, better yet, concurrently), an attempt should be made to return to the source of the wastes to determine whether any reductions can be achieved. If no acceptable disposal solutions exist, source reduction is the most feasible waste management technique for a large waste volume, especially in small nations. In many countries, landfill disposal will be the only option for disposing of massive amounts of hazardous waste.

7.7 TYPES OF LANDFILLS

In practice, three distinct possibilities have been observed: Among the historical sites that are no longer open are the following: The amount of information available about the inputs, historical experience, when the site was closed (those closed after the early 1960s are more likely to have issues), and what restoration and aftercare measures, if any, were implemented all have an impact on the management of any issues associated with these. Historical significance and continued use: The amount of information available about past inputs, when licensing/authorization mechanisms allowing for more robust controls and management regimes were implemented, whether before or after the early 1960s, changes in the type of inputs, and so on, all have an impact on the most effective problem management (Issaka& Ashraf, 2017).

The following are examples of high-potential locations: These are configured to handle the waste and will be subject to appropriate license/authorization conditions, producing were appropriate for phased restoration and operating under quality-assured management regimes, ensuring that operational, restoration and aftercare issues are kept to a minimum.

Sites of Landfill Diverse Operational Environments Containment These include those that provide material containment and are commonly found in naturally occurring clay deposits, those that involve filling void spaces created by clay extraction for brick or cement manufacturing, and those that supplement or replace natural containment with engineered liners. Liquid collecting devices are typically installed beneath and above the liners in all of these cases to collect any leachate that seeps through or is stored on them.

Those that permit a delayed release. The unsaturated zone formed by the underlying natural or engineered strata allows for attenuated physical, chemical, and biochemical release processes, lowering the pollution potential of any leachate and ensuring that the recipient environment's capacity to accommodate any released species is not exceeded. There are parallels here with the recipient-capacity environment's ability to absorb material fluxes caused by natural mineral weathering and water movement.

A specially engineered landfill, for this study, is designed to provide containment and includes elements such as installing drainage to recover any leakage, as well as leachate management, which may include recirculation and gas control systems. Working with a cell system that gradually recovers almost always results in success (Ighodaro, Lategan, & Yusuf, 2013).

Containment is accomplished by (a) selecting a suitable location, (b) adhering to stringent operational criteria, and (c) carefully selecting and pre-treating wastes to ensure long-term environmental reduction (depending on their chemical properties) (based on their chemical characteristics).

7.8 A SPECIALLY DESIGNED LANDFILL

However, this type of landfill, which allows for the controlled disposal of garbage on land, is frequently the only viable option for hazardous wastes. In theory, and for a limited time, a landfill site can be made ecologically safe by using the proper location, methods, and management. The landfill must be adequately prepared, administered, and managed to reduce risks to human health and the environment. The same techniques should be used for site selection, design and construction, operation and monitoring, closure, and post-closure care (Haregeweyn et al., 2015).

7.9 HIGHWAY CORRIDORS, PARKING LOTS, AREAS OF HEAVY TRAFFIC

The metropolitan environment is densely packed with abandoned autos. Numerous automobiles and trucks are strewn over highways and parking lots, and several major thoroughfares are gridlocked. Every day, we rely on our automobiles to get from point A to point B. While a large number of automobiles makes travel easier, they also offer a significant danger of contamination due to emissions and fluid leakage (Ha, Van Dung & Ngoc, 2013). When oil or petroleum spills occur on roads or parking lots, neighboring soils can get polluted, resulting in elevated lead levels.

Figure 7.7: Areas of heavy traffic.

Source: https://upload.wikimedia.org/wikipedia/commons/8/8c/Motorcycles_on_Civic_Boulevard_20080918.jpg

As technology advances and people become increasingly reliant on automobiles for daily travel, roads are becoming more prevalent around the world. Highways in the United States cover more than a fifth of the entire land area of the country, totaling more than 4 million miles. These motorways have transformed the terrain irreversibly and pose several risks to animals. Apart from wreaking havoc on the environment, roads can affect population dynamics and ecology. Roads are a hot area of study in conservation biology and environmental science due to the variety of ecological consequences they may have (Guerra et al., 2017).

7.10 POLLUTION

Additionally, highways might be utilized to transfer pollutants into the environment. The metamorphosis of wood frogs has been expedited by road tire particles. Deicing salts that leak from roadways can be harmful to neighbor ponds' wood frogs and spotted salamanders. Bone deformities in frogs living near streets are one of the most likely outcomes of pollution. It has been established that deicing salts change frog behavior and impair their capacity to move. This may impair their capacity to acquire food and dodge predators. Additionally, if oil from automobiles that travel on roadways enters the environment, it might kill wildlife.

Additionally, roadside light and noise pollution can be detrimental to the health of drivers and pedestrians. Automobile noise impairs birds' auditory transmission and warning signals, resulting in a decline in bird populations near roadways. While not all birds are impacted equally, those whose song frequencies correspond to those of cars are more likely to be missing from roadside locations (Frankl et al., 2016). Road noise can alter the makeup of bird communities by eradicating some species and lowering bird populations. Roads can also interfere with frogs' cries, making it more difficult for them to find a partner.

Highway lights may affect animals that rely on light to regulate their biological activity. If the sun is not out, robins may mistakenly sing their morning songs during the night. Additionally, road lights can change bats' flying paths. Certain frog species may take longer to metamorphose when exposed to artificial lighting at night. Hatchling sea turtles that rely on light to find their way to the ocean may be confused by traffic lights, leading them away from the water and onto the road. It is possible that they will perish from starvation, predators, or passing automobiles and will never reach the river (García-Ruiz et al., 2017). When nesting females travel to the beach to lay their eggs, road lights may confuse them, making their return to the sea more difficult. In general, road development facilitates the spread of alien species. Cane toads have grown highly invasive in Australia as a result of their need for roadside areas for mobility. The roadside mounds constructed by invading fire ants provide them with a haven. Numerous invasive plants use spreading to new locations as a common tactic. Roads can assist the spread of invasive species by lowering the number of species competing for newly cleared areas.

Automobile pollution has a detrimental effect on the air, land, and water quality. When the ozone layer, which protects the Earth from damaging ultraviolet radiation from the sun, is depleted as a result of nitrous oxide emissions, humans suffer. Sulfur dioxide and nitrogen dioxide combine to generate acid rain in the presence of rain, which is damaging to both plants and structures due to the damage it causes to crops, trees, and other forms of flora. Near highways, oil and gasoline spills seep into the earth, while wasteed fuel and particles from automobile exhaust contaminate lakes, rivers, and wetlands.

7.10.1 Automobile Pollution

Vehicle and truck owners can take a range of measures to mitigate the environmental impact of automobile pollution. Compared to older, less-maintained automobiles, electric, hybrid, and other clean, fuel-efficient vehicles have a lesser environmental impact. Before making a purchase, research the vehicle's fuel economy and environmental rating. Increased pollution levels imply a clean environment. Remove extraneous objects like roof racks to avoid abrupt acceleration and deceleration. While routine automobile maintenance, such as oil changes and tire rotations, is critical, you should also avoid driving as much as possible. Whenever possible, walk, ride, or use public transportation.

Environmental pollution is impacted by several variables, including the quantity and distribution of pollution sources within a specific region, the terrain of the area, the climate conditions, and the kind of pollutant (Fenta et al., 2021). Industrialization and traffic congestion are the two primary elements that influence the degree of pollution in a given location. Automobile exhaust contains not only carbon monoxide and nitrogen oxides, but also incomplete combustion of hydrocarbons present in oil and its derivatives, which are presumably less harmful than the first two (milder irritants) but form highly toxic and carcinogenic PAH chemicals. Automobile exhaust may include benzopyrene, a known carcinogen. Polycyclic aromatic hydrocarbons become more hydrophobic as their molecular weight increases, making them nonpolar and neutral organic compounds. They are made up of two or more condensed benzene rings that have been linked in a variety of ways. Even though solid clay particles are insoluble in water, they rapidly attach to and link up with PAHs in animal fat tissues. Nearly all PAHs have been linked to cancer. Natural sources of PAHs include forest fires and volcanic eruptions. Additionally, automotive exhaust and hazardous waste sites contribute significantly to pollution. PAHs provide the strongest soil particle connections.

Figure 7.8: Oil spill on a car.

Source:https://www.dreamstime.com/oil-spill-car-image194591344

Certain PAHs found in soil may represent a threat to underground water sources. When benzene, a recognized carcinogen, and naphthalene, a water-soluble solvent, come into contact with water, tar (which also contains PAHs) dissolves them. When exposed to sunlight and other chemical molecules, PAHs can degrade into airborne chemicals over days or weeks. Bacteria are primarily responsible for soil and water breakdown, which can take months or even years. By interacting with the air and fuel, which play a significant role in the creation of PAHs, catalysts can significantly reduce the concentration of PAHs in exhaust gases. PAH levels in internal combustion engines vary according to operation temperature, charge, fuel quality, and speed. The highest amounts of PAH emissions are seen in urban areas with high population densities, intensive traffic, and short-distance automobile trips. Increased emissions can be a result of engine breakdowns and wear-and-tear (Damaneh et al., 2022).

According to a previous study, these chemicals' emissions might be decreased by up to 90%. Diesel engine catalysts also contribute to the reduction of overall PAH emissions, but to a lesser extent than gasoline engine catalysts. Gasoline with a reduced PAH content can help to reduce PAH emissions from the exhaust. Diesel automobiles emit more PAHs than gasoline-powered vehicles. PAH can be degraded or absorbed by bioremediation, which employs fungi and bacteria, or phytoremediation. The purpose of this research is to determine the effect of polycyclic aromatic hydrocarbons (PAHs) on the soil along the M-21 major highway, which runs through the National Park "Fruka Gora."

While the number of polycyclic aromatic hydrocarbons (PAHs) in soil decreases with depth, some, such as phenanthrene, may be detected at depths as little as 2 m, and their abundance is directly related to biodegradation, which is influenced by external influences. Certain PAHs may be absorbed and degraded by plants.

The majority of PAHs are found in soil, while a few are also identified in plant roots. The leaves of Asteraceae plants have the greatest amounts. In both depths, twenty-four PAH components were detected at levels below the detection limit of 2 ppm or the practical quantification limit of 5 ppm (0-20 cm and 20-40 cm). Numerous PAHs, including phenanthrene, benzo(a), benzo(b), benzo(h), benzo(a,h), dibenzo(a,h), and indenopyrene, were analyzed based on their spatial distribution and depth in the sample. In sites 3, 4, 5, and 6, a total of nine PAHs were detected, with two detected in the top layer (0-20 cm) and eight detected at a depth of 20-40 cm, with two detected in the top layer (0-20 cm). Numerous locations include the greatest levels of benzo(a)anthracene and benzo(b)fluoranthene. According to the study, between six and sixteen distinct PAHs have been detected in Poland's contaminated and unpolluted regions. Their concentrations varied from 1.5 to 30.1 g/kg in uncontaminated soils but reached 33100 g/kg in heavily contaminated soils near petrochemical facilities. Benzo(a)pyrene concentrations on a route with an average daily traffic density of 8000 cars were found to range between 7.1 and 10.6 grams per kilogram. According to the findings of this investigation, benzo(a)pyrene concentrations were between 5.8 and 6 g/kg in a variety of areas with comparable traffic densities. Utilizing a route that has not been traveled in at least four decades resulted in some unexpected findings at Locality 7 (Blake et al., 2018).

At further depths, a significant quantity of PAHs was detected. This is due to a large number of diesel-powered big transport vehicles and a low frequency of automobiles. Fluoranthene, benzo(a)anthracene, chrysene, benzo(b), benzo(k), dibenzo(a,h)anthracene, and indeno(1,2,3-c,d)pyrene were detected in the greatest concentrations in a single soil sample (12, 65, 8.9, 65, 18, 23, and 83 g/kg, respectively). Agricultural soils with a depth range of 0 to 20 cm were evaluated for PAH concentration in three distinct sites (industrial districts, city centers, and suburbs). According to a study, industrial regions contained the highest concentrations of congeners. Benzo(a)anthracene, benzo(b)fluoranthene, benzo(k)fluoranthene, and indeno(1,2,3-c,d)pyrene concentrations in these samples varied between 17.2 and 14.2 ng g-1 (Cogo, 2022). Due to the absence of additional sources of PAHs in the vicinity, it seems as though PAHs are created in the soil sample site. The kind of vehicle, the status of the engine, and the surrounding environment all affect the amount of pollution emitted. Automobiles driven by gasoline emit fluoranthene and pyrene, whereas diesel-powered cars emit naphthalene and acenaphthene, both of which are PAHs. The concentrations of naphthalene, fluorine, anthracene and acenaphthylene are all below the MDL or PQL in all locations, demonstrating the extensive usage of gasoline-powered automobiles in these regions.

According to a past study, total PAH levels in soil can increase in industrial and agricultural regions, as well as along major roadways. Endogenous PAH concentrations in soil and plants vary between 1 and 10 g/kg and 10 to 20 g/kg, respectively, according to Edwards. Endogenous PAHs are produced by plant synthesis, forest and prairie fires, volcanic activity, and other natural activities. During an investigation, PAHs were discovered in concentrations ranging from 10 g/kg (at site 1, 20-40 cm) to 306.9 g/kg (in total) (locality 4, 20-40 cm). Total PAH concentrations in surface soil (0–5 cm) samples taken from various functional areas in Ji'nan, Shandong Province, China, ranged from 1.31 mg kg1 (dry weight) to 254.08 mg kg1 (wet weight), with an average total PAH content of 23.25 mg kg1. Industrial regions had the greatest amounts, reaching up to 53.97 mg kg1 along major highways (Cerda et al., 2018).

7.10.2 Household Sites

Although homes may be a low-risk source of soil pollution, this is not the case. Soil contamination, on the other hand, might occur during the construction of a residence. The usage of petroleum products in construction trucks has the potential to cause leaks. There is a danger that paint will leak.

Roof-related fibers have the potential to blow down and damage soil life. These are only a few examples of soil pollution caused by construction on construction sites.

Figure 7.9: Household site soil pollution.

Source: https://commons.wikimedia.org/wiki/File:42.4_kg_of_food_found_in_New_Zealand_household_rubbish_bins.jpg

Furthermore, homeowners may unwittingly harm their soil. The main source of worry is the overuse of pesticides and herbicides. Chemically treated wood is occasionally used in landscaping. If not used properly, this can harm the soil, plants, and the microbial life it supports. Additionally, too much winter salt might be harmful to your soils.

The soil was once a pristine material that enveloped the earth. However, in rare cases, people have mistakenly or purposely sprayed harmful substances on it. Waste can harm the environment as well as the health of humans, plants, and animals (Chuchkalov, Fadeev, & Alekseev, 2020).

The largest risk of soil contamination is seen in urban areas and old industrial sites. A soil test should be performed if you are worried about the safety of the soil surrounding your home or property. Indeed, the vast majority of soil is perfectly safe for play, gardening, and recreation, but it's always better to be cautious than sorry!

Pesticides, petroleum products, radon, asbestos, lead, chromate copper arsenate, and creosote are some of the most common contaminants identified in urban soils. Human activities are the primary cause of urban soil contamination. Manufacturing/industrial waste disposal, land development, local rubbish disposal, and excessive pesticide or fertilizer usage are a few examples. Even large-scale activity, such as autos and trucks, can damage the soil; have you ever noticed a shimmering pool in your driveway behind your car? That is oil, a petroleum product, and some of it may wash into the soil when it rains!

The source and amount of pollution deposited in soils have a significant impact on how contamination spreads over an area. The type of soil also has an impact on its distribution. Certain poisons, for example, may be absorbed more quickly into groundwater sources when kept in the sand rather than clay. This is due to the higher rate of penetration of coarse-grained sandy soil types (Zhang & Huang, 2015). Contaminants firmly maintained in fine-grained clay soils or organic material in surface soils will accumulate if left undisturbed (e.g., no excavation or tillage) (e.g. no excavation or tillage).

Individuals might be exposed to soil contaminants in a variety of ways. The most well-liked are as follows:

- Soil ingestion
- Volatile and dust inhalation

Consuming contaminated soil-grown food and absorbing it through the skin

Urban gardens are usually a fantastic idea, but you must first understand your soil. Many vegetables and plants can absorb contaminants as they grow. If you eat them, you are putting your health in danger. Vegetables and plants may also show signs of dust from filthy soil. Contaminants may be consumed if they are not thoroughly cleansed. Some garden beds may also have chemically treated wood linings. If you did not design your garden beds, you must analyze your soil since pesticides may have leaked into the soil.

Garden soil contamination might be difficult to detect. Soil contamination, on the other hand, might occur during the construction of a residence. Petroleum products are leaking onto construction trucks. There is a danger that paint will leak. Roofing fibers, for example, might fly into the soil and devastate the environment. These are only a few examples of soil pollution caused by construction on construction sites (Yakupoglu, Rodrigo-Comino &Cerdà, 2019).

Homeowners may inadvertently injure their soil. The main source of worry is the overuse of pesticides and herbicides. Chemically treated wood, on the other hand, is used in landscaping on occasion. If not used correctly, this can be harmful to the soil as well as the plant and microbial life that it supports.

Many vegetable plots might be found on old landfills and garbage grounds. If you have a garden on one of them, you have very certainly uncovered ancient pottery and other detritus while digging in it. This is true in a lot of cities. Cities have grown by filling up gaps with garbage, debris, and soil. Former landfills are frequently converted into gardens. This is a place where people may get away from the hustle and bustle of the city and enjoy some fresh air. Many of these gardens, on the other hand, have survived and provided a peaceful respite for thousands of garden owners. This is a rare occurrence in large cities that should be preserved.

Garbage and rubble at landfills are not a major problem. The soil gathered from excavation sites isn't any better. The source of the problem is industrial and chemical waste. We now know a lot more about soil degradation in gardens that have been created in abandoned landfills. Some environmental protection officials studied soil and discovered contaminants in far higher concentrations than what should be in the soil (Van Straalen, 2002). Because there is little evidence on the long-term effects, and because humans are exposed to toxins from a variety of sources, specific treatments are recommended. We don't know what the long-term effects of working in polluted soil or eating fruits and vegetables grown in polluted soil are.

As a result, the instructions from experts are simple. If these recommendations are followed, the negative consequences of contaminated soil can be avoided.

Formerly active landfills, gasworks, and other polluting enterprises can contaminate the environment. Some landfills have been around since the 1800s. Because the garbage in these dumps was often scattered out at random, some areas were exceedingly poisonous while others were not. Heavy metals, petroleum, and tars were once discovered in industrial waste dumps. Because most oil has degraded, only trace amounts of oil may be found in soil (Webb, Herrick, &Duniway, 2014). Tars degrade slowly and can remain in the soil for decades. Tar has been found in several allotment plots. Heavy metals such as lead and cadmium do not break down and hence remain in the soil. Pollutants do not evaporate as a result. Contaminants have been found on the surface as well as up to a half-meter depth. The ground is not stable and is constantly being mixed as you dig. This implies that even if you sprinkle compost on top of the soil, the topsoil will almost certainly become contaminated.

Pollution is something that no one can completely avoid. Many of the poisons found in soil can also be found in other parts of our environment, such as food, cigarette smoke, paint, textiles, and plastics, as well as in the air. Science has improved our ability to analyze dangerous chemicals and determine if they are the cause of sickness or other negative outcomes. It is difficult to predict the long-term effects of low-level hazardous chemical exposure on humans. As a result, extreme caution should be exercised while handling contaminated soil, particularly with little infants who frequently get soil on their hands and in their mouths. Fruits and vegetables may cause problems if they are not cleaned or peeled before consumption. Although the precise mechanism by which pollutants enter the body is unknown, we do know that some are absorbed more than others. Vegetables should be grown on the uncontaminated ground wherever possible. It is critical to emphasize that youngsters should avoid eating vegetables grown in polluted soil owing to their increased sensitivity (Van Leeuwen et al., 2019).

Following the recommendations in this pamphlet will help to reduce the risks in the long term. Lead, cadmium, and tars are the most common pollutants found in gardens developed on historical waste sites. In the past, accidents and negligence with these chemicals, such as bad working conditions, were known to cause injury. There has been no increase in illness among those who live on polluted land. This does not, however, rule out the possibility of negative consequences that are undetected by current analytical approaches. It can be difficult to separate the effects of unclean soil from the effects of other sorts of pollution or toxins to which we are exposed in our daily lives. According to one research, little children who live on polluted ground may have tiny amounts of harmful compounds in their systems as a result of putting soil in their mouths when playing outside. Soil quality guidelines are developed by officials from environmental and health groups, and they contain a maximum amount of contaminants that may be allowed in the soil. These limitations are determined with a safety margin based on how much dirt children swallow, how much they receive from other paths, and how much contamination they can endure. Soil pollution levels at historic dump sites are far over legal limits in some locations. Evaluations are necessary for these environments to determine what should be done to avoid potential harm.

7.11 THE MOST EFFECTIVE GARDENING METHOD

Vegetables may be grown in raised beds with clean soil. Vegetables can be grown in plant pots or bags of clean soil, or raised beds constructed of uncontaminated soil. In its most basic form, a raised bed is a wooden structure filled with about half a meter of soil. A protective net should be erected at the bottom of the raised bed to prevent the new clean soil from getting polluted with the underlying dirt. Both rain and worms may pass through this net. The remainder of the garden should be filled with earth (Tsoraeva et al., 2021). This reduces the likelihood of children accumulating soil on their hands and feet and helps to prevent the spread of dangerous soil and dust. To hide dirty garden soil, grow grass or put tiles where there will be a lot of foot traffic. Flower beds can be planted with plants that cover exposed land, or the soil can be covered with mulching, compost, bark, or other materials, especially if children play in the yard.

Figure: 7.10: Growing vegetables in raised bed gardens.

Source: https://upload.wikimedia.org/wikipedia/commons/8/89/Hochbeet_Vorderm%C3%BChle_1.JPG

7.12 FORMER FARMLAND WITH THE BUILD-UP OF CONTAMINANTS

Between 1910 and 1950, lead-arsenate was used in the manufacture of insecticides in the United States. It wasn't until decades later that specialists and farmers realized lead was dangerous. As a result, lead has been discovered in the soil of abandoned farms. Herbicides have been developed and commercialized since the 1950s. These chemicals must be used correctly to avoid harming the soil, plants, and even humans. High-load fertilizer applications can cause soil contamination depending on the crop and the type of fertilizer used (Tully et al., 2015).

Lead is a persistent hazardous pollutant of concern due to its role in the debate over the impact of human activity on the environment. The goal of this review is to explain how the lead is absorbed by plants and to connect lead absorption to plant toxicity. Our research is also focused on the processes by which plants or plant systems detoxify lead. Because of its distinct physicochemical properties, lead is a highly sought-after heavy metal. Since the beginning of time, humans have used lead. Industrialization, urbanization, mining, and other human activities have spread lead from the earth's crust into the soil and environment. Lead forms a variety of complexes with various soil constituents in the soil, and only a minuscule fraction of this complexed lead is phytoavailable. Even though lead is not required for plant growth, roots absorb it from the soil solution and thus enter the food chain. Roots absorb lead via the apoplastic route or Ca2+-permeable channels.

When it comes to the behavior of lead in soil and the absorption of this heavy metal by plants, several factors come into play, including the type of lead, particle size, and cation exchange capacity, surface area of the roots, root exudation, and degree of mycorrhizal transpiration. Because Casparian strips surround the endodermis, lead is mostly accumulated in root cells after absorption. The negatively charged cell walls of roots also act as a lead trap. Excessive lead deposition in plant tissue harms several morphological, physiological, and biochemical processes. This has several negative consequences, either directly or indirectly (Telles, Guimarães, &Dechen, 2011). Phytotoxicity is achieved by altering the permeability of the cell membrane, interacting with enzymes involved in plant metabolism, reacting with the phosphate groups of ADP or ATP, and substituting for essential ions. Lead poisoning impairs ATP synthesis, lipid peroxidation, and DNA damage by producing an excessive amount of reactive oxygen species (ROS). Lead not only inhibits seed germination and root extension, but also reduces plant growth, transpiration, chlorophyll production, and the amount of water and protein a plant can absorb.

Due to a confluence of factors, lead has an even greater impact on plant development than it already does. This includes a decrease in magnesium and iron absorption as well as an increase in CO2 deficit due to stomatal closure, all of which worsen the effects of lead on plant growth. Plants use a variety of defensive mechanisms to combat lead exposure. Examples of these strategies include the development of complexes to sequester lead in vacuoles, the binding of lead to phytochelatins, glutathione, and amino acids, and the production of osmolytes. A secondary defense option for combating lead-induced ROS production is to activate various antioxidants.

The duration of pesticides is determined by soil chemistry and microbial activity. The content of organic matter, sand, silt, and clay in the soil affects the texture, which in turn affects the composition. Pesticide persistence may be affected by soil pH, an important chemical characteristic. There are two microbiological characteristics of the soil environment: species and abundance.

Herbicide effectiveness and persistence are affected by soil-herbicide binding (adsorption), leaching, and vapor escape, respectively (volatilization). Heavy clay, organic matter, or both soils have a higher potential for herbicide carryover due to increased herbicide binding to soil particles. The "tie-up" method should be avoided to limit herbicidal activity. Herbicide stocks are being hoarded, which could harm sensitive crops in the future.

Herbicides have the greatest potential to adhere or remain in soils with more than 3% organic matter for fragile rotation crops. Herbicide absorption and carryover are less likely in coarse to medium-textured soils with low organic matter concentrations (less than 3%), which are more common in Pennsylvania soils. Herbicide carryover, on the other hand, can occur in any soil type under the right conditions (Salesa &Cerdà, 2020).

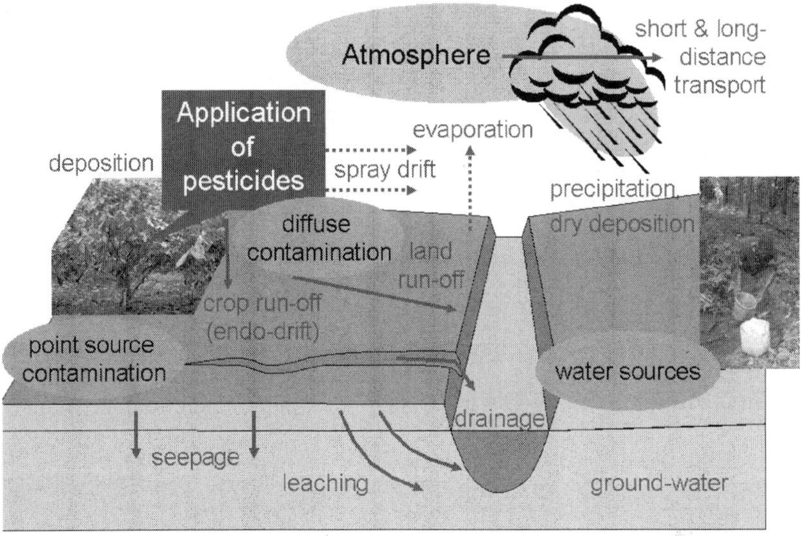

Figure 7.11: Herbicide/pesticide pollution.

Source: https://upload.wikimedia.org/wikipedia/commons/4/47/Env_contamination1.if.gif

Some herbicides, most notably triazines and sulfonylureas, can be affected by soil pH. Herbicides are degraded chemically and microbiologically in soil. Soils with higher pH take longer to deteriorate chemically and microbiologically. When soil pH rises, the chemical breakdown of triazine and sulfonylurea herbicides is reduced, especially at pH 7.0. Herbicides are more easily absorbed by plants in higher-pH soils because they are less connected to soil particles. Herbicides like triazine and sulfonylurea have a longer active life in soils with higher pH, making them more available to plant life. Certain herbicides do not endure or carryover, regardless of whether the soil is acidic or alkaline (Rodrigo-Comino, 2018).

Low pH can also harm triazine and sulfonylurea herbicides. Both pesticide families evaporate faster in soils with a pH less than 6.0. Herbicides such as atrazine become chemically bonded to soil particles in acidic soils, rendering them ineffective for weed control. To achieve the desired results with these two herbicide families, acidic soil must be limed.

Imidazolinone herbicides such as imazaquin (Scepter) and imazethapyr (Imaze) are more resistant to low soil pH (Shaner, 2017). When the pH of the soil falls below 6.0, imazaquin and imazethapyr are gradually bonded to or absorbed by soil particles. Soil bacteria, which are the primary decomposers, are less likely to be exposed to herbicides that have been adsorbed. Even though soils with lower pH have a higher absorption capacity, the herbicide can be released months later, making it available to plants and potentially harming a sensitive follow crop.

Herbicides are most likely degraded primarily by soil microorganisms. Microorganisms must be present in sufficient numbers to allow the decomposition process to proceed as quickly as possible (fungi, bacteria, protozoans, and others). Certain environmental conditions must exist for microorganisms to efficiently produce pesticides. Moisture, temperature, pH, oxygen, and mineral nutrition all have an impact on microbial activity. The ideal conditions for microbial growth and herbicide breakdown are frequently found in warm, well-aerated, rich soil with a pH close to neutral. The chemical properties of an herbicide influence its persistence. These characteristics include the molecule's solubility in water, vapor pressure, and susceptibility to chemical or microbiological change or destruction.

CHAPTER 8

MOST POLLUTED SOIL IN THE WORLD

CONTENTS

8.1 Introduction 196

8.2 Soil Pollution 197

8.3 Agricultural Areas 200

8.4 Polluted Soil in Industrial Areas 204

8.5 Most Polluted Soil By Radioactive Elements 208

8.6 Conclusion 213

8.1 INTRODUCTION

Soil comes from decomposed organic matter, which means it was once living. Soil is a complex mixture of solid and liquid particles, air, water, and microorganisms. The major component of soil is solid material. Rock fragments with grains ranging in size from clay to gravel. Soil is a living organism itself that is made up of billions of tiny organisms called microorganisms or microbes. These microbes help to break down the materials in the soil into other nutrients that are used by plants and other organisms. Soils are formed when rock or organic matter weather over long periods (Beilicci, Beilicci& Stefănescu, 2014). Because this process is so slow, many soil layers are mostly undisturbed. Disturbance results in the introduction of new rock, erosion of old rock, and the mixing of sediments with the soil layers. These processes often result in a vertical layering effect. When the different soil layers are exposed to varying amounts of water, the resultant soils can be very distinctive from each other. This represents an important opportunity for biotic and human change.

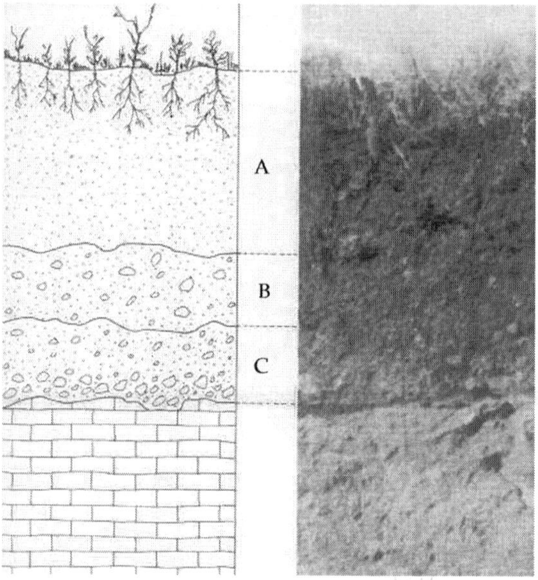

Figure 8.1: A, B, and C represent the soil profile, a notation firstly coined by Vasily Dokuchaev (1846–1903), the father of pedology. Here, A is the top-soil; B is a regolith; C is a saprolite ; the bottom-most layer represents the bed-rock.

Source: https://commons.wikimedia.org/wiki/File:Estructura-suelo.jpg

Today, the soil becomes an endangered resource. Mineral nutrients such as organic matter, nitrogen, phosphorus, potassium, and calcium are in rapid depletion. Soil contains essential minerals and nutrients needed for plants to grow and as a result, it acts not only as a habitat but also as a food source. With soils in rapid erosion, we are sure to lose our ability to grow crops, which will put an end to life on earth.

Soil is the base of all life on Earth. Everything we eat and use, everything we rely on, every living thing that has ever existed, all of it depends on soil. Soil is more than just dirt; it is a complex ecosystem of living organisms, water and gases, mineral particles, and organic matter. And not all soil is the same. Soil can be sandy or sticky, moist or dry, dark or light-colored. Different plants grow best in different kinds of soils. Most of the food we eat comes from plants that grow in the soil. The ground is where plants get water and nutrients to help them grow strong and healthy. Soil also holds onto carbon dioxide gas; without it, the Earth would be too warm for most plants and animals to live comfortably (Rodrigo-Comino et al., 2018). But much of our topsoil is being eroded or degraded every day by human activity like deforestation and farming, as well as by climate change and other environmental factors like drought. When topsoil gets eroded or polluted, it can't support plant life the way it should, which means less food for us and a greater chance that we'll find those chemicals in our bodies too. Soil also plays a critical role in fighting climate change: Healthy soil helps store carbon dioxide, while unhealthy soil releases it back into the atmosphere. So the healthier we can keep our soil, the better we can help combat climate change something more important than ever before.

8.2 SOIL POLLUTION

Soil pollution is the contamination of the soil with harmful chemicals and materials. For example, when a car runs over a puddle, it can splash toxic chemicals onto the ground, polluting the soil in that area.

Figure 8.2: An illustration of soil pollution.

Source: Suh J, Kim S-M, Yi H, Choi Y. An Overview of GIS-Based Modeling and Assessment of Mining-Induced Hazards: Soil, Water, and Forest. International Journal of Environmental Research and Public Health. 2017; 14(12):1463. https://doi.org/10.3390/ijerph14121463

The most common cause of soil pollution is acid rain. This occurs when certain gases, such as sulfur dioxide and nitrogen oxide, are released into the atmosphere. When these gases encounter water molecules in the air, they form acids that eventually fall back to Earth's surface in the form of acid rain. This acid rain can then cause soil pollution by releasing metals like mercury and lead into the soil. Soil pollution may also occur when human activities release toxic substances into the environment. For example, oil spills can contaminate soil and groundwater, making them unsafe for human consumption. The disposal of industrial waste products can also cause soil pollution if it is not properly contained or disposed of (Blake et al., 2021).

Mining does also lead to soil pollution. Some of the minerals mined from the lower layers of the earth's surface may be harmful to the upper layer of the earth's surface; and if not handled with care while being extracted from the earth, they may spill and end up contaminated the topsoil; affecting plant life, and some in serious cases, affecting animal life for animals which reside on the top layers of the earth's surface. Construction activities do also lead to soil pollution in that; after construction activities are completed, most construction sites are not cleared of all the toxic waste which was being used for construction purposes. The decay of such toxic waste leads to the contamination of the soil. Waste disposal mechanisms lead to soil pollution in different ways; for instance, the dumping of car batteries contaminates the soil for the produce toxic chemicals which make the soil.

Soil pollutants are divided into different categories. There are biological agents; agricultural practices; radioactive pollutants; urban waste; and industrial waste. Biological agents include the likes of organic manure and digested sludge from living organisms such as the likes of human beings, birds, and other animals. Agricultural practices, as the name suggests, include activities such as the application of pesticides, fertilizers, and herbicides among other activities. Radioactive soil pollutants include substances that are radioactive such as the likes of radium, thorium, uranium, and nitrogen, among others (Rickson, 2014). Urban waste soil pollutants are the day-to-day waste materials that are accumulated due to human consumption of different materials; and include the likes of garbage waste and rubbish waste. Industrial waste, as the name suggests, includes the soil pollutants which are released from industries; and they include the likes of steel, pesticides, textiles, drugs, glass, cement, and petroleum among other waste materials.

Soil pollution is harmful to the life of all living organisms, both plants, and animals. The inhalation of gases emitted from contaminated soils may cause several health issues to human beings; such as headaches, nausea, fatigue, skin rash, eye irritation, and extreme exposure rates lead to kidney damage, liver damage and cancerous diseases.

8.2.1 Most Polluted Soils in the World

The most polluted soils in the world are found in regions that are highly exposed to soil pollutants. Those regions include agricultural areas; industrial areas; waste disposal areas, and mining regions among others.

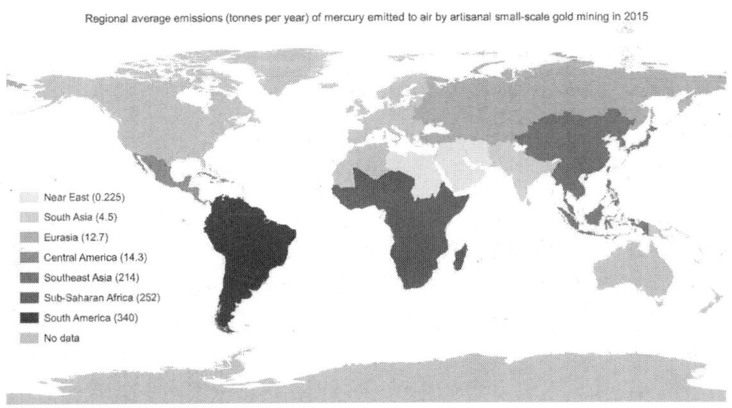

Figure 8.3: An illustration of the world's regions with the most polluted soil.

Source: https://www.fao.org/3/cb4894en/online/src/html/chapter-03-5.html

8.3 AGRICULTURAL AREAS

Creating implies cultivating and agriculture is the reason people can live within the world today. It could be a fundamental implies of survival, without which there would be starvations all over the world. For thousands of a long time, agribusiness was a normal handle that did not hurt the arrival was done on. In reality, agriculturists were able to pass down their land for numerous eras, and it would still be rich as ever. However, present-day agrarian hones have begun the method of rural contamination. This preparation causes the corruption of the environment, land, and environment due to the modern-day by-products of horticulture.

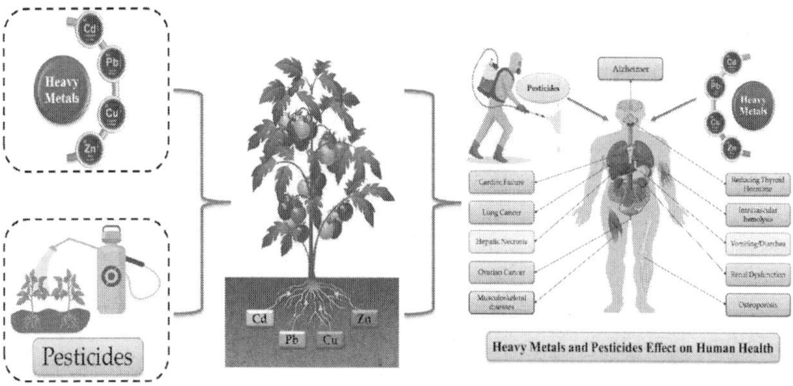

Figure 8.4: An illustration of pesticide application, which leads to soil pollution.

Source: Alengebawy A, Abdelkhalek ST, Qureshi SR, Wang M-Q. Heavy Metals and Pesticides Toxicity in Agricultural Soil and Plants: Ecological Risks and Human Health Implications. Toxics. 2021; 9(3):42. https://doi.org/10.3390/toxics9030042

To start with, the earliest source of contamination has been pesticides and fertilizers. Modern-day pesticides and fertilizers got to bargain with the nearby insects that have existed hundreds of times alongside the invasive species. And so, they are loaded with chemicals that are not found in nature.

Once they have been splashed, it does not vanish totally. A few of it blends with the water and leaks into the ground. The rest is ingested by the plant itself. As a result, the nearby streams that are provided water from the ground get to be polluted, as do the creatures that eat these crops and plants (Kowalska et al., 2018).

Polluted water utilized for the water system is one source of contamination. Much of the water we utilize comes from groundwater stores, rivers, and rains. Whereas majority of it is clean and unadulterated water, other sources are contaminated with natural compounds and heavy metals. This happens due to the transfer of mechanical and agrarian waste in neighborhood bodies of water. As a result, the crops are uncovered to water, which has little sums of mercury, arsenic, lead, and cadmium broken up in it. The method of agrarian contamination gets to be harder to battle when such water harms animals and causes edit disappointment.

Further issues are caused by soil disintegration and sedimentation. The soil is comprised of numerous layers, and it is as it were the highest layer that can bolster cultivating or touching. Due to wasteful cultivating practices, this soil is cleared out open for disintegration and leads to declining richness each year. Whether disintegrated by water or wind, all this soil must be kept someplace or the other. The sedimentation causes the soil to construct up in zones such as streams, streams, ditches, and encompassing areas. And so, the method of rural contamination avoids the normal development of water, oceanic creatures, and supplements to other prolific regions (Poesen, 2018).

Creating exceptional crops and diminishing the common species in a certain zone has gotten to be the standard for cultivation. In any case, it is counting on the strategy of rural pollution. With the section of cutting-edge crops, the nearby people ought to bargain with advanced diseases, pests, and weeds that it isn't competent of battling. As a result, the invasive species obliterate the adjacent vegetation and characteristic life, changing the natural framework for all time. Regularly especially the case with genetically engineered foods, which make plant and animal species that can wipe out the existing species in a matter of a long time.

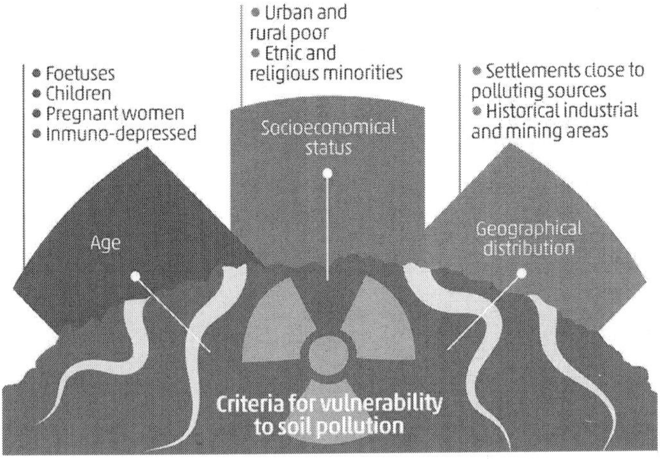

Figure 8.5: A summary of the causes and effects of agricultural soil pollution.

Source:https://www.fao.org/3/cb4894en/online/src/html/chapter-04-4.html

The use of fertilizers, manure, and other natural wastes containing heavy metals such as arsenic, cadmium, mercury, and lead can too lead to a collection of these heavy metals within the soil. Cultivating strategies like water system can too lead to an amassing of selenium. When these substances are washed into conduits or filter into groundwater sources or get ingested by plants and are in the long run consumed by creatures and people influencing their health. Heavy metals can harm animals from polluted water or nourishment.

Serious cultivating operations significantly contribute to soil disintegration and sedimentation as millions of rich soils are broken down using stormwater runoffs, which ends up collecting as silt in waterways, streams, lakes, seas, or other land districts. Hence, it influences water quality by making it messy or polluting it with the agrochemical buildups displayed within the soils. Sedimentation moreover contributes to the build-up of the rural toxins in conduits and other land zones. Sedimentation may too limit light entrance in water, in this manner influencing marine life (Shaner, 2017).

Manures and Biosolids regularly contain nutrients, like nitrogen, carbon, and phosphorus. Moreover, since they are mechanically prepared, they may too have inside them contaminants such as individual care items and pharmaceuticals. These items have been found in human and animal bodies and are accepted to have negative health impacts on natural life, creatures, and humans. Agricultural contamination gets to be indeed harder to oversee with such sorts of natural contaminants (Pijl et al., 2020).

Bad land administration to leads to an irreversible decrease in soil richness. Significant land administration is significant for keeping agrarian contamination to the least level. Hence ranchers ought to have the mindfulness of how their activities can affect the environment.

The manure and fertilizers as a rule contain overabundance chemical supplements, particularly phosphorus and nitrogen, and cause supplement contamination from rural sources. Abundance supplements can have appalling results on water quality and the survival of marine life. When these supplements are washed into the water frameworks, for example; waterways, lakes, streams, or seas amid stormy periods, it changes the marine and freshwater nutrient cycles and as a result the species composition of the particular environments. The foremost common result is eutrophication, which depletes the water oxygen, and in result, can kill oceanic life.

The soil pollution caused by several agricultural activities causes several negative effects on the environment, to both plant life and animal life. Agrarian contamination is the main source of contamination in water and lakes. Chemicals from fertilizers and pesticides make their way into the groundwater that closes up in drinking water. Health-related problems may happen because it contributes to a blue baby syndrome which causes death in newborn children. Oil, degreasing operators, metals and poisons from gear causes health issues when they get into drinking water.

Fertilizers, manure, waste, and smelling salts turn into nitrate and phosphates, and when washed into adjacent water bodies, the generation of algaegets upgraded that diminishes the sum of oxygen show in water, which comes about within the passing of numerous marine animals. Again, microscopic organisms and parasites from creature waste can get into drinking water, which can posture genuine health risks for different marine life and creatures. Hence, the oxygen levels are likely to decay, which can cause the passing of angles and other water creatures.

Eutrophication is the thick development of plant life and algae on the water surface, causing algal blooms. In case of intemperate use of fertilizers and pesticides, nitrogen, phosphorus, and other chemical supplements get washed into adjacent surface waters by rain or water system and lead to the eutrophication of waterways and lakes by supporting the generation of algae (Pham et al., 2018). Eutrophication affects oxygen breakdown in the water, which can negatively affect the marine ecosystem by harming marine biotas.

It is additionally connected to an increased risk of poisoned fish harming people, leading to death.

The unrestrained use of fertilizers and pesticides combined with other agrochemicals control weeds, and help improve yields. Be that as it may, the positive impacts of these substances are finite since the soil is likely to drain them into the environment causing chemical pollution. When they stay in the soil for a longer time period, yields are decreased, and the soil loses the ideal qualities to produce healthy crops. They have the potential of polluting waters and plants and harming soil microorganisms as well as useful insects.

The chemicals that constitute pesticides and other diverse sorts of agrochemicals can cause long-lasting harm to the soil. This may continuously change the soil microbial activities and soil chemistry and decrease soil fertility. Thus, each year millions of fertile soils are affected due to the use of manufactured fertilizers, pesticides, and herbicides combined with other farm practices.

The environment is often delicate to little changes which can lead to enormous impacts within the biological system. The constant use of chemical products in agrarian processes debases the soils and waters, affects plant and animal life, continuously modifying the biological systems which support biodiversity. Furthermore, the use of pesticides can harm useful insects, and soil microorganisms. In case insects disappear from the environment, plants will be negatively impacted as these insects are used in the fertilization of crops.

Water contamination is a major issue caused by agrarian chemicals. Improper irrigation methods and water systems primarily lead to water contamination from surface runoff, both to surface and groundwater (Pietro & Mahajan, 2022). Excessive use of fertilizers and pesticides, and destructive substances reach our lakes, streams and certainly the groundwater leading to broad pollution of channels and ground waters and devaluing water quality. Soil disintegration and sedimentation similarly debase the water, making it muddled, and sullied. In turn, plants, people, creatures are adversely influenced since we require clean drinking water to live and remain healthy.

8.4 POLLUTED SOIL IN INDUSTRIAL AREAS

With the coming of the Industrial Revolution, people were able to advance into the twenty-first century. Advancement grew rapidly as a result of science and technology, and the manufacturing age came. With all of these came one more effect, industrial pollution. Earlier, businesses were small mechanical offices that only produced smoke as the primary toxin. In any case, since the number of production plants were limited and operated only a certain number of hours a day, the levels of pollution caused was not great. But when these production lines later became full-scale businesses and manufacturing units, the issue of industrial pollution started to gain more centrality.

Figure 8.6: Domestic waste/mixed waste site.

Source: https://upload.wikimedia.org/wikipedia/commons/9/97/Payatas-Dumpsite_Manila_Philippines02.jpg

The issue of industrial pollution has become serious for organizations endeavoring to battle against environmental degradation. Industrial pollution takes on different faces but the most polluted soil in the world are found in industrial areas (Priori et al., 2020). The need for successful environmental protection has led many governments to push for bylaws on industrial pollution control. In most cities, impromptu developments occur wherein companies disregard rules and standards and contaminate the environment in large scale. Moreover, many small-scale manufacturers don't have sufficient capital and depend on government awards to run their day-to-day businesses, frequently eluding environmental guidelines and discharge an expansive amount of harmful gasses into the environment. The discharged compounds can cause soil and water contamination when spilled on the earth.

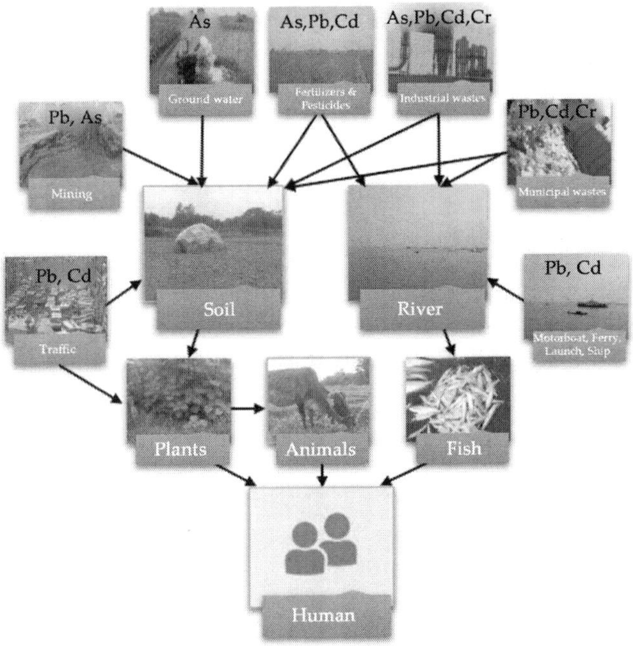

Figure 8.7: An illustration of the list of industrial soil pollution causes.

Source: Islam MM, Karim MR, Zheng X, Li X. Heavy Metal and Metalloid Pollution of Soil, Water and Foods in Bangladesh: A Critical Review. International Journal of Environmental Research and Public Health. 2018; 15(12):2825. https://doi.org/10.3390/ijerph15122825

The most polluted soil due to industrial pollution affects the environment in different ways. The impacts of mechanical contamination are far-reaching and at risk to influence the biological system for numerous a long time to come. Most businesses require expansive sums of water for their work. When included in an arrangement of forms, the water comes into contact with heavy metals, harmful chemicals, radioactive waste, and natural sludge. These are either dumped into open seas or waterways (Pimentel & Burgess, 2013). As a result, numerous of our water sources have a tall sum of mechanical waste in them, which truly impacts the health of our biological system. The same water is at that point utilized by ranchers for water system purposes, which influences the quality of nourishment that's produced. Water contamination has as of now rendered numerous groundwater assets futile for people and natural life. It can at best be reused to encourage utilization in businesses.

Soil contamination is making issues in agribusiness and wrecking nearby vegetation. It moreover causes unremitting health issues to the individuals that come in contact with such soil on a day-by-day basis. Air contamination has driven a soak increment in different ailments, and it proceeds to influence us on a day-by-day premise. With so numerous little, mid and expansive scale businesses coming up, discussing contamination has taken a toll on the health of the individuals and the environment. The issue of industrial pollution shows up as it causes characteristic rhythms and plans to drop level, meaning that the common life is getting affected truly. Regions are being lost, species are getting to be wrapped up, and it is harder for the environment to recover from each characteristic catastrophe. Major calamities like oil spills, fires, the spillage of radioactive materials, and hurt to the property are harder to clean up as they are bringing about negative influence in a shorter period.

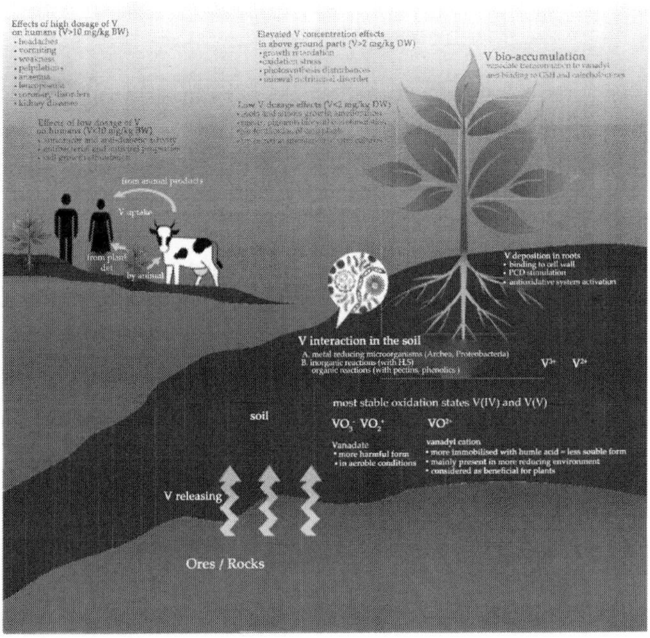

Figure 8.8: An illustration of some of the effects of industrial soil pollution, i.e, loss of plant life.

Source: Hanus-Fajerska E, Wiszniewska A, Kamińska I. A Dual Role of Vanadium in Environmental Systems—Beneficial and Detrimental Effects on Terrestrial Plants and Humans. Plants. 2021; 10(6):1110. https://doi.org/10.3390/plants10061110

With the rise in mechanical contamination, global warming has been expanding at a relentless pace. Smoke and greenhouse gasses are being discharged by industries into the discussion, which causes an increment in worldwide warming. Melting of ice sheets, termination of polar bears, surges, tsunamis, storms are a few of the impacts of global warming. Industrial contamination proceeds to cause noteworthy harm to the soil and all of its tenants due to chemical wastes, pesticides, radioactive materials, etc. It influences natural life and environments and disturbs normal environments. Animals are getting to be terminated, and territories are being destroyed. The expanding fluid, strong and dangerous wastes weaken environmental health and affect nourishment, water, and health security. Mechanical contamination calamities, oil spills and radioactive spillage, take a long time to decades to clean up (Nearing, Lane, & Lopes, 2017).

Cadmium improvement of soil can be related to mechanical contamination. Topsoils polluted by mine drains had a wide range of cadmium concentrations. Industrial effluents are commonly released to surface water seepage frameworks after clarification in the following lakes. Later examinations have uncovered high concentrations of cadmium within the overbank and foot silt of the streams.

8.5 MOST POLLUTED SOIL BY RADIOACTIVE ELEMENTS

Radioactivity could be a marvel related to unsteady nuclear cores with an abundance of vitality or mass, which suddenly break down radiating ionizing radiation by electromagnetic waves, gamma beams, or streams of subatomic; alpha, beta, or neutron particles. The action of a specific radioactive substance is characterized by the consistent decay rate and the half-life, the time taken for the movement of a given amount of a radioactive substance to decay to half of its initial valueee, and it could be that ten half-lives are required for each radioisotope to be disposed of. Since the half-lives of different nuclei change from seconds to billions of a long time, the time required for to decay.

Decay chains

The radionuclides of natural decay chains are ubiquitous in rocks and soils. In nuclear science, a decay chain refers to a sequence of radioactive decays of various radioactive decay products as a set of successive transformations. It is also known as the “radioactive cascade”. The most common types of radioactivity are alpha-decay, beta-decay, gamma-emission, positron emission, and electron capture. Radioactive decay constitutes the emission of particles and/or energy as one atom transforms into another(Rao, 2014). In alpha (α) decay, helium ions are released from the nucleus. In beta (beta) decay, neutrons in the nucleus are converted into protons and electrons.

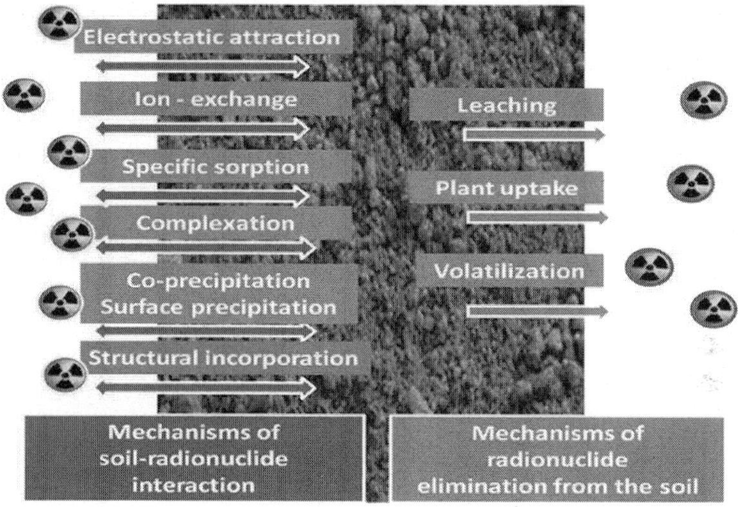

Figure 8.9: An illustration of a brief summary of the interaction between soil and radioactive elements.

Source: Smičiklas, I., & Šljivić-Ivanović, M. (2016). Radioactive Contamination of the Soil: Assessments of Pollutants Mobility with Implication to Remediation Strategies. In M. L. Larramendy, & S. Soloneski (Eds.), Soil Contamination - Current Consequences and Further Solutions. IntechOpen. https://doi.org/10.5772/64735

A few radionuclides occur actually within the environment, and their nearness is either cosmogenic or earthly. The 3H, 7,10Be, 14C, 26Al, and 39Ar are the most radionuclides delivered after the interaction of air gasses with enormous beams. On the other hand, the rocks, minerals, and subsequently the soil, contain naturally occurring radioactive materials, characterized by long half-life periods.

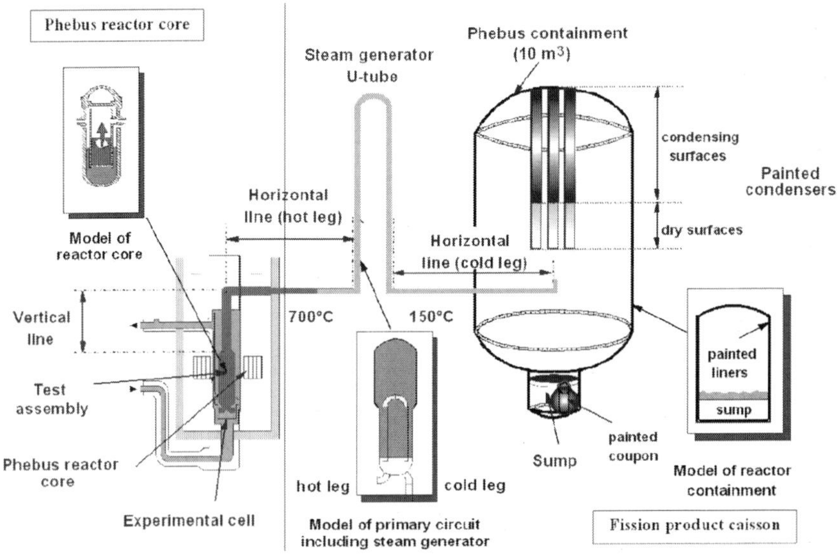

Figure 8.10: How radionuclides are produced during the fission process.

Source: Flores y Flores A, Ferretto D, Marková T, Mazzini G. Analysis of Release Model Effect in the Transport of Fission Products Simulating the FPT3 Test Using MELCOR 2.1 and MELCOR 2.2. Sustainability. 2021; 13(14):7964. https://doi.org/10.3390/su13147964

The term radioactive contamination means the exposure of radioactive substances to solids, liquids, gasses, or biota. Radioactive contamination occurs through the discharge of man-made radionuclides, making ionizing radiation one of the common natural components that cause soil corruption (Nigussie et al., 2015).

Indeed even though the radioactive pollution of the environment is generally uncommon, it requires an extraordinary consideration because of the major corrupting impacts of ionizing radiation on living tissues. The negative impacts are in relation to the amount of retained health, the impact of radiation is degeneration of the affected cells. In earthly biological systems, soil compares to the major receiving pool of radiated radionuclides.

Figure 8.11: An illustration of the radioactive soil contamination.

Source: https://www.123rf.com/stock-photo/radioactive_pollution.html

Processing of radioactive waste may result in a coincidental discharge of the radionuclides during the characterization, division, transportation, treatment, and exchange stages. To be specific, within the middle of atomic reactor operation, most metallic surfaces oxidize and form a layer of the crumbling film that oxides with various components. Depending on the composition of the reactor materials and their support components, temperature control and shutdown period, the deterioration things and their relative degrees may differ.

Another key source of soil pollution with radionuclides is atomic weapons tests, especially air. Barometric explosions create radioactive flotsam and jetsam of diverse molecules, which are divided within the troposphere and stratosphere and accelerate over some minutes to 1 year or longer. Atomic mishaps are the events that cause major damage, such as the ones in Chernobyl, in Ukraine, 1986 and Fukushima, in Japan, 2011. These two occasions caused global pollution of the environment. In the aftermath, radioactive pollution within the soil in 1993 was found to be 3500 times higher than the level before the Chernobyl accident.

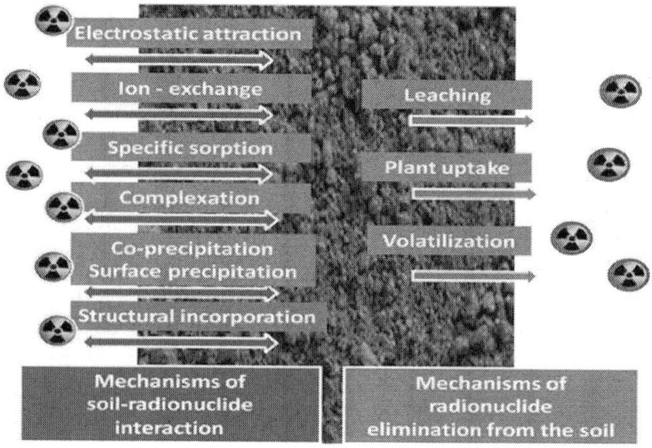

Figure 8.12: An image showing the effects of soil pollution by radioactive elements.

Source: https://www.intechopen.com/chapters/52054

Chemical extraction is the method that promotes the redistribution of contaminants from the radioactive spill. The polluted soil must be unearthed and treated off-site. After the treatment, the soil is returned to its unique area, whereas the movement remains concentrated within the extraction medium. The soil remnant is hence treated to accelerate the action and return the leaching reagents to the process. Plants can also be grown to promote phytoextraction.

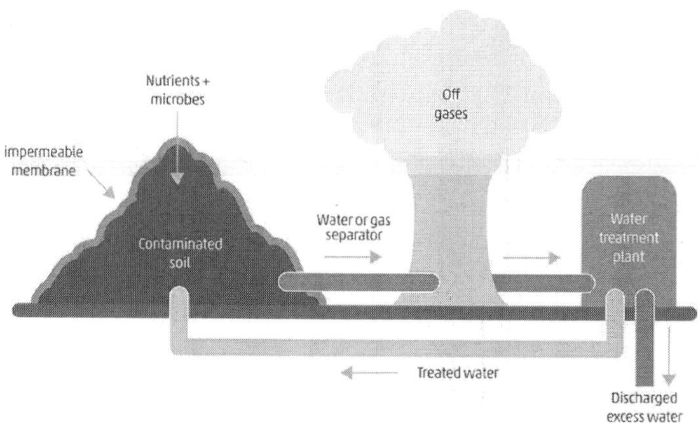

Figure 8.13: An illustration of chemical extraction.

Source:https://www.fao.org/3/cb4894en/online/src/html/chapter-13-4.html

To remove the target metal from the soil, the force of the radionuclide-chelating operator complex must overcome the force of the bonds keeping radionuclide joined to the soil surface. The proficiency of EDTA is prevalent, and it is more often than not connected at pH 4, as the EDTA-complexes can be re-adsorbed on soil surface locations at lower pH. Chemical extraction forms have an extensive potential in the restoration of the soil that have experienced radioactive pollution and their viability can be improved by optimizing reagent types and concentration, soil ratio, pH, mixing, and other variables.

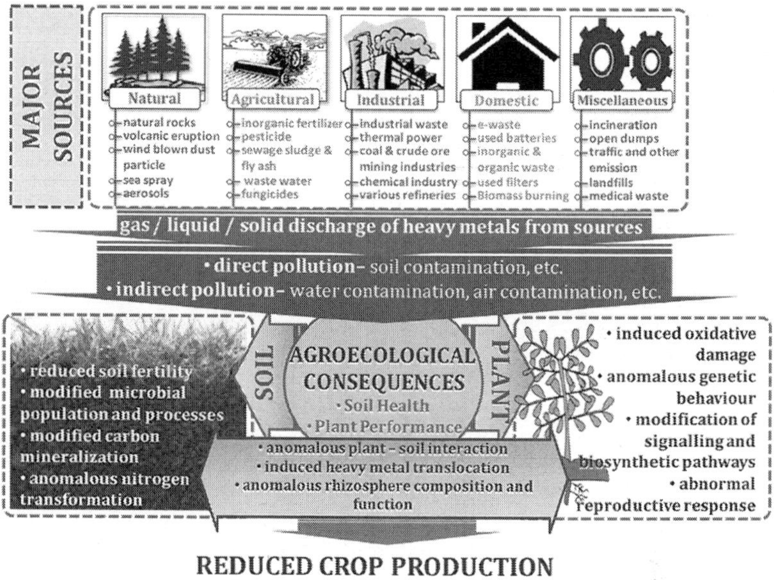

Figure 8.14: An illustration of the agroecological impacts of soil pollution.

Source: Srivastava V, Sarkar A, Singh S, Singh P, de Araujo ASF and Singh RP (2017) Agroecological Responses of Heavy Metal Pollution with Special Emphasis on Soil Health and Plant Performances. Front. Environ. Sci. 5:64. doi: 10.3389/fenvs.2017.00064

8.6 CONCLUSION

A wide set of soil and common parameters affect the radionuclide activity in soil systems. Weaker bonds between the compound and soil components include higher adaptability of damageto the biota. Understanding of its characteristic behavior by choosing disposal is of basic centrality for ideal remediation advances (Lefèvre et al., 2021). Due to the colossal number of variables that affect the result of the soil recovery, the course of activity must be done on a case-by-case present.

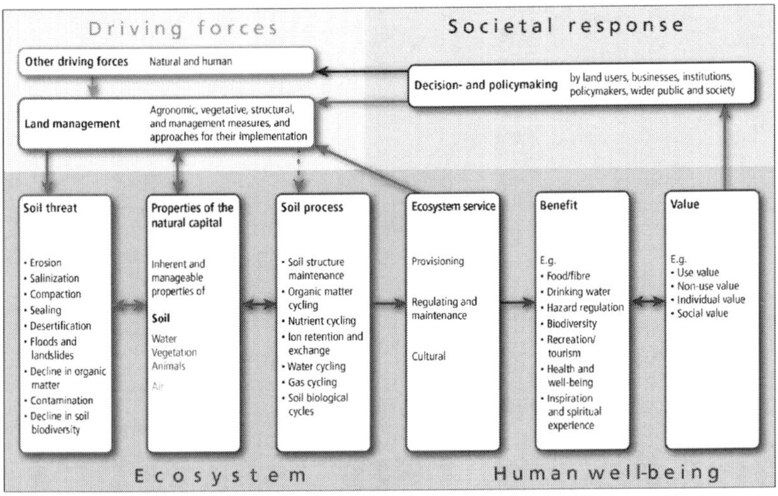

Figure 8.15: An image showing a brief summary of the possible solutions to the most polluted soil in the world.

Source: Schwilch G, Lemann T, Berglund Ö, Camarotto C, Cerdà A, Daliakopoulos IN, Kohnová S, Krzeminska D, Marañón T, Rietra R, Siebielec G, Thorsson J, Tibbett M, Valente S, Van Delden H, Van den Akker J, Verzandvoort S, Vrînceanu NO, Zoumides C, Hessel R. Assessing Impacts of Soil Management Measures on Ecosystem Services. Sustainability. 2018; 10(12):4416. https://doi.org/10.3390/su10124416

CHAPTER 9

RISK ASSESSMENTS AND THE REMEDIATION OF CONTAMINATED SOILS

CONTENTS

9.1 Introduction 216

9.2 Bioremediation 217

9.3 Long-Term Development and Restorative Land Use in Underprivileged Regions 220

9.4 The Size and Environmental Impact of Bioremediation Soil Pollution in Disadvantaged Communities 221

9.5 How Soil Characteristics Influence Biodegradation 222

9.6 Environmental and Human Health Risks of Contaminated Soils 224

9.7 Toxaphene 224

9.8 Toxic Heavy Metals 226

9.9 Bio-Stimulant 229

9.10 Whey 232

9.11 Bioaugmentation 235

9.12 Eco-Friendly Technologies 237

9.13 Technology 241

9.14 Environmental Science and Technology 242

9.15 Conclusions 243

9.1 INTRODUCTION

The problem of soil pollution is a difficult one that must be addressed. To live, we must all understand the importance of soil. If we perceive soil contamination as an issue, we can address it more promptly. Every stakeholder has a role to play, from people to the government. Below are some remedial techniques after soil pollution(Ettler, 2016).

Ecosystem services are the cornerstone of all human civilizations and international economies. The long-term management of ecosystems has an impact on the resilience of social-ecological systems. As a result, the health of the environment, human health, and food security are all compromised. This causes an annual economic loss in the millions of dollars. Habitat loss and fragmentation, unsustainable harvesting, pollution, climate change, and the introduction of alien invading species have all harmed more than 60% of the world's ecosystems. As a result of industrial activities, toxic substances are discharged into the environment (Keesstra et al., 2018). Toxic toxins build up in the biosphere because most ecosystems are incapable of breaking them down. The most frequent soil contaminants are halogenated hydrocarbons, heavy metals, and petroleum products. Contamination from industrial or agricultural activity may endanger both animals and humans, as well as impact economic systems by rendering land unfit for cultivation and other uses.

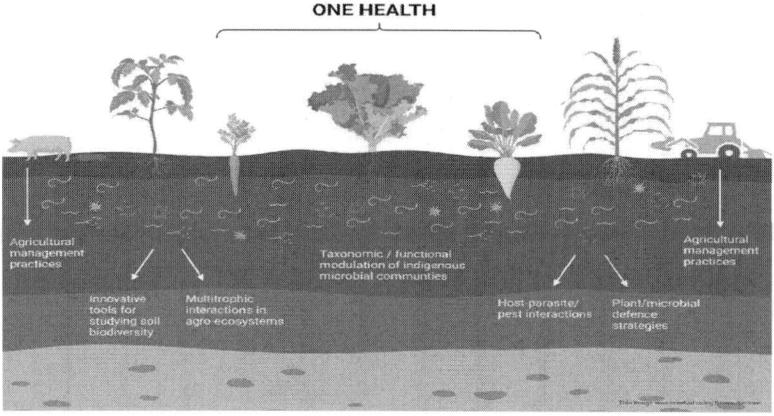

Figure 9.1: Soil amendments for immobilization.

Source: https://www.mdpi.com/journal/sustainability/special_issues/soil_microbiomes_agriculture

9.2 BIOREMEDIATION

In bioremediation, enzymes derived from living organisms are utilized to break down, remove, or immobilize dangerous substances. Bioremediation was utilized to clean up the Exxon Valdez oil spill on Prince William Sound's beach, although on a far bigger scale than is customary for bioremediation. Ex-situ soil bioremediation occurs when contaminated soil is physically removed from its original position, whereas in situ soil bioremediation occurs when the soil remains in its original location. Because of the necessity for transport, ex-situ bioremediation may be more energy and cost-intensive than in situ bioremediation. Bioremediation can be a more cost-effective alternative to typical chemical cleaning (Kelly et al., 2020). Bioremediation, according to some experts, will continue to gain popularity as new technologies overcome current limitations. Researchers in bioremediation encounter a variety of obstacles, including the following: Increasing the bioavailability of hydrophobic and recalcitrant compounds, mapping the complex anaerobic/aerobic metabolic pathways of contaminants, and improving microbial transport to reach contaminants are just a few of the goals we're pursuing as we learn more about the soil microbial community and how it changes during bioremediation processes. If emerging technologies can overcome these restrictions, bioremediation may become a more feasible alternative for soil remediation.

9.2.1 What Is Bioremediation and How Is It Performed?

Bioremediation involves the cultivation of microorganisms that absorb pollutants such as oil, solvents, and pesticides to get food and energy. These microbes may also transform harmless gases like carbon dioxide into water. The proper mix of temperature, nutrition, and food is required for bioremediation (Kheyrodin, 2016). If these ingredients are absent, contamination cleanup may take longer. Introducing "amendments" to the environment, such as molasses, vegetable oil, or plain air, can assist ameliorate circumstances that aren't conducive to bio-remediation. As a result of these adjustments, the bioremediation process will be hastened.

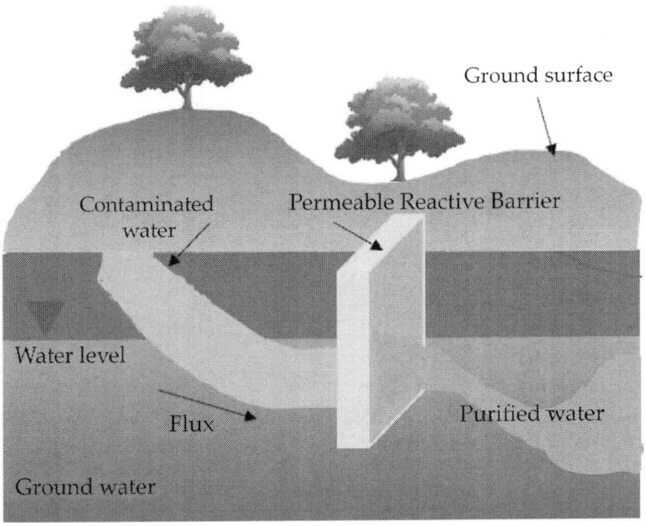

Figure 9.2: Soil Bioremediation.

Source:https://encyclopedia.mdpi.cn/Fh71BLgk44U-u2OQ34df2rYe0Xb4

This can be done "in situ," or at the source of the pollution, or "ex-situ," or in a location other than the source of the pollution. When the soil is too rich or the temperature is too low for microbial activity to thrive, ex-situ bioremediation may be required. Excavating and cleaning the soil above ground will account for a considerable percentage of the expense of an ex-situ bioremediation project. Toxin concentration, temperature, soil density, and whether bioremediation will take place in situ or ex situ all influence how long the bioremediation process takes.

9.2.2 The Benefits of Bioremediation

Using bioremediation instead of traditional cleaning procedures has several advantages. Bioremediation, a practice that uses chemicals and microorganisms to remove toxins from groundwater and soil, is a typical approach to environmental protection. As a result, bioremediation has a smaller environmental effect than other techniques of cleaning (Kumarasinghe, 2021).

During the bioremediation process, one or two potentially harmful metabolites are produced (mainly because contaminants and pollutants are converted into water and harmless gases like carbon dioxide).

Bioremediation may be a better option if you don't have a lot of money to invest in machinery or labor. The United States Environmental Protection Agency (EPA) launched 1,507 bioremediation projects in 2018.

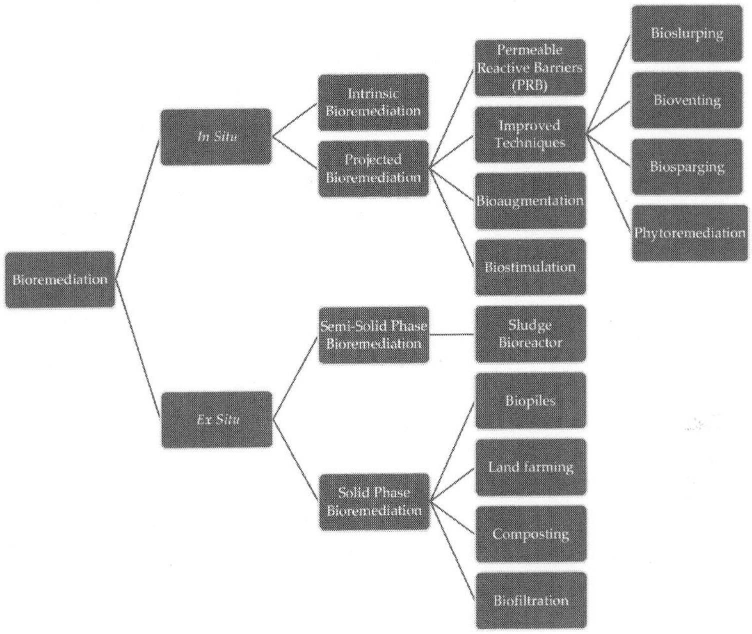

Figure 9.3: The process of waste bioremediation.

Source: Sales da Silva IG, Gomes de Almeida FC, Padilha da Rocha e Silva NM, Casazza AA, Converti A, Asfora Sarubbo L. Soil Bioremediation: Overview of Technologies and Trends. Energies. 2020; 13(18):4664. https://doi.org/10.3390/en13184664

In 1989, the oil tanker Exxon Valdez went aground off the coast of Alaska, spewing an estimated 11 million gallons of oil into the environment. Bioremediation, on the other hand, was gaining traction as a viable approach to oil spill cleanup. Exxon Mobil Corporation (XOM) and the Environmental Protection Agency (EPA) both started researching alternative chemicals. At first sight, the bioremediation effectiveness research appeared promising.

During 1989 and 1990, around 2000 applications of fertilizer were applied to the impacted regions, totaling approximately 100,000 pounds of fertilizer. By the middle of 1992, when the cleaning procedure was deemed complete, almost all of the oil compounds had been dissolved by fertilizer (Kavian et al., 2019).

9.3 LONG-TERM DEVELOPMENT AND RESTORATIVE LAND USE IN UNDERPRIVILEGED REGIONS

Traditional soil remediation approaches include energy-intensive and soil-disturbing equipment such as soil extraction and landfill disposal, which has been a hot topic for decades.

Because of a lack of economic incentives, governance, knowledge, and skill, soil contamination in marginalized areas, such as rural areas in developing countries or sparsely populated regions in developed countries, is largely ignored. Certain places are frequently undervalued by governments and investors. The primary focus of this research is on underserved areas where economic incentives for soil remediation of any kind are lacking.

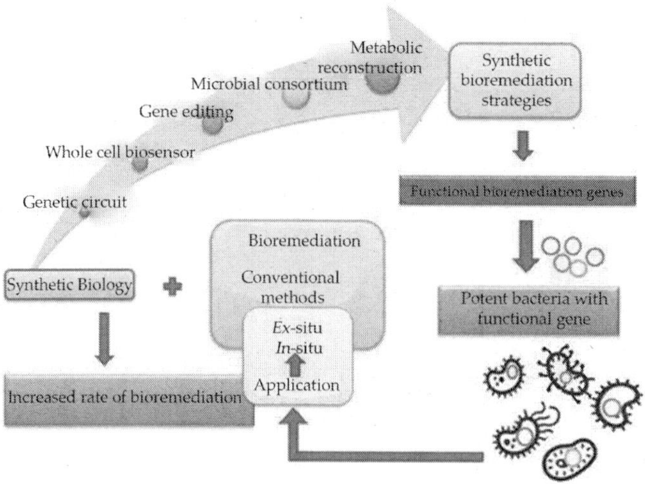

Figure 9.4: Alternative Strategies for Microbial Remediation of Pollutants.

Source: https://www.frontiersin.org/files/Articles/522718/fmicb-11-00808-HTML-r1/image_m/fmicb-11-00808-g001.jpg

Although there are fewer severe time constraints than in densely populated urban areas, energy and resource efficiency remains a concern. Poor countries have fewer restrictions on the use of hazardous compounds, and the extent of soil contamination is often unknown, implying that a wide range of activities may occur unintentionally on damaged land. Because only 56% of Nicaraguan communities have regular garbage collection services and 87% of the country's waste dumps are unlicensed, soil contamination occurs both accidentally and on purpose in Nicaragua, the country where the experimental investigation described in this thesis was carried out (Jomaa et al., 2013). Environmental concerns are frequently overlooked in Nicaragua and other impoverished countries as a result of the country's economic difficulties. Many important polluted sites in emerging economies will need to be cleaned up in the future. Resource-related issues may also affect sparsely populated regions in developing countries. In northern Sweden's Jamtland County, there are 3,100 potentially polluted areas, 359 of which have been contaminated by fuel handling and contain oil depots.

Even though it was once common practice to replenish the engine oil of forest machinery, it is unknown how frequently small leaks from forestry machinery and other sources occur. Investing in soil remediation is difficult in Jamtland because the rural property market is sometimes too small to meet regulatory requirements. Traditional rehabilitation is prohibitively expensive for contaminated sites of this magnitude, so they receive less attention than they require. Traditional soil remediation methods, such as soil extraction and landfill disposal, can be costly or time-consuming to implement. As a result, due to the enormous costs associated with removing harmful compounds from polluted soils, remediation may be impossible in places of minimal economic importance, such as developing countries and sparsely populated regions in industrialized countries. Organic additives from Mid Sweden University (MIUN), such as whey and the fermented whey, have been shown to accelerate the natural breakdown of specific chemicals in diesel fuel. In situ whey application is a low-priority in situ bioremediation method based on readily available by-products. It is an energy-efficient and essentially non-disruptive method.

9.4 THE SIZE AND ENVIRONMENTAL IMPACT OF BIOREMEDIATION SOIL POLLUTION IN DISADVANTAGED COMMUNITIES

Industrialization has wreaked havoc on the earth's soil over the last two centuries, contaminating at least one-third of the planet's ecosystems. In addition to animal and environmental harm, soil contamination may endanger human health. The most common pollutants in European soil are metals such as lead and mercury, as well as mineral oil. Approximately 250,000 of Europe's three million potentially contaminated sites may need to be cleaned up right away. Although the extent of soil pollution in developing countries is unknown, widespread use of persistent pesticides is having an impact on large areas of land.

The three basic methods through which soil pollutants reach a human's body are ingestion, inhalation, and skin absorption (Issaka& Ashraf, 2017). Soil contamination may be absorbed in two ways: by eating plants grown on contaminated soil and then consumed by people or livestock reared for human consumption, or by directly consuming contaminated soil in its raw form (geophagia). As a result, children under the age of three are particularly vulnerable to this type of soil pollution exposure. Because of the danger of breathing particles that eventually enter the circulation, field workers are most at risk of inhalation-related disease. Dermal absorption (through skin contact) is a severe issue for volatile organic chemicals, whereas non-volatile soil pollutants such as heavy metals provide only a minor danger.

When a pollutant enters the environment, its "environmental destiny" refers to the life cycle of the contaminant. Extrinsic factors like oxygen availability, temperature, and sun exposure all influence a contaminant's environmental destiny and persistence. Organic pollutants can be completely reduced to carbon dioxide and water in certain situations, or they might undergo a partial degradation process that leads to the formation of hazardous or harmless metabolites. Heavy metals and several persistent organic contaminants can stay in the environment for a long period in their natural state. A pollutant's decomposition in the natural environment is influenced by a variety of elements, including whether it is organic, inorganic, or quickly degradable. Organic pollutant breakdown rates can be influenced by a variety of environmental conditions. The context for the research's experimental features, tropical clay soils deficient in nutrients, has two essential deciding variables: the availability of nutrients and the physicochemical properties of the contaminated soil.

9.5 HOW SOIL CHARACTERISTICS INFLUENCE BIODEGRADATION

Clay soils are less likely to support soil life because their fine texture renders them unsuitable for the majority of microorganisms. Because of the roughness of the soil, the mass transfer may be a barrier to biodegradation in thick soils. Gases have a difficult time diffusing through compacted, moist soil, resulting in inadequate soil air circulation and, as a result, a lack of aeration. Non-polar chemicals, such as PAH, are less accessible due to their affinity for organic molecules and clay particles (Ighodaro, Lategan, & Yusuf, 2013). Microbial activity is most concentrated in the higher layers of the soil, where the majority of organic matter and oxygen are present because here is where the majority of microbial activity occurs. Earthworms and other macrofauna are often useful in the degradation process.

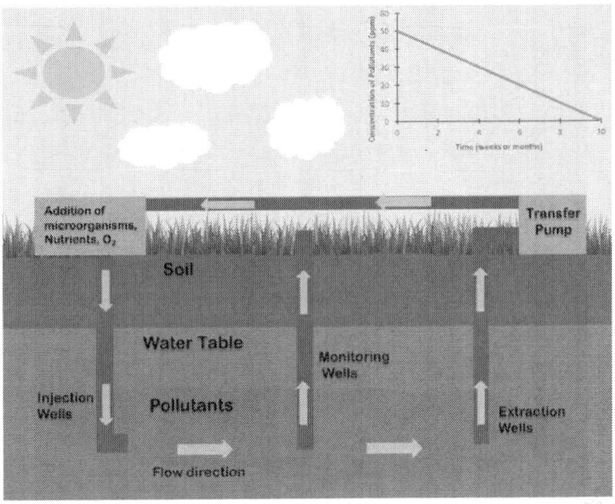

Figure 9.5: *Schematic representation of the different steps involved in biodegradation.*

Source: By Hoodlind - Own work, CC BY-SA 4.0, https://commons.wikimedia.org/w/index.php?curid=64647624

9.5.1 Nutrient Accessibility and Biodegradation

When physicochemical parameters such as temperature, pH, and pollutant bioavailability can accommodate bacterial activity and degradation, chronically low nitrogen and phosphorus levels in the soil are the limiting factor for organic pollutant breakdown. Injecting organic by-products can provide a variety of nutrients that can accelerate degradation in a variety of ways. Polluted soil, on the other hand, has a significant amount of degradable organic carbon from fuel. Organic chemicals that help in macronutrient breakdown include peptone, calcium lactate, yeast extracts, nicotinamides, riboflavin, pyridoxine, thiamine, and ascorbic acid. These chemicals, however, impede down degradation in large quantities (Haregeweyn et al., 2015). We investigated several contaminants, each with its unique features and degradation mechanisms, in this study: This group includes diesel fuel hydrocarbons, persistent organic pollutants, and heavy metals.

9.6 ENVIRONMENTAL AND HUMAN HEALTH RISKS OF CONTAMINATED SOILS

9.6.1 Diesel Engines

Because petroleum hydrocarbons are complex combinations of aliphatic and aromatic hydrocarbons, the vast majority of them are biodegradable. The bulk of diesel fuel's composition is made up of diesel range organics (DRO), alkylbenzenes, naphthalenes, and trace quantities of polycyclic aromatic hydrocarbons (PAH). It may be feasible to leverage the molecules' high intrinsic energy content to accelerate the microbial breakdown process. Because of the complexity of diesel's chemical makeup, it is usually essential to utilize a mix of microorganisms capable of breaking down a variety of distinct molecular structures to break it down. Microorganisms produce a plethora of polar compounds, most of which are harmless, to degrade fuel. However, because of their water solubility, the more mobile polar compounds are, the higher the potential of groundwater pollution. In rare circumstances, such microbial transformation has been demonstrated to increase toxicity, particularly in oxygen-depleted conditions.

For a long time, POPs (permanent organic pollutants) were referred to as environmental contaminants (POP). POPs are compounds that are hydrophobic, lipophilic, and very stable. POPs enter the environment through several routes, including chemical plant output and leakage from a wide range of consumer products. Agriculture pesticides are another important source of POP pollution. The United Nations Environment Program (UNEP) has identified twelve persistent organic pollutants (POPs) that must be addressed promptly. All nine POPs are insecticides (Ha, Van Dung & Ngoc, 2013). In agriculture, insecticides are commonly employed in non-targeted ways. Only 2% of the pesticide sprayed was identified on the leaf surface in one test, implying that the remaining 98 percent of the pesticide did not reach its intended target and instead ended up in the soil and air.

9.7 TOXAPHENE

Between 1950 and the early 1990s, the bulk of toxaphene usage occurred in the United States, the ex-Soviet Union, Central America, and Brazil. Toxaphene contamination was first identified in Canada and Scandinavia in the early 1980s. Toxaphene was discovered in lipid-rich polar fish and mammals in these nations, even though they are hundreds of kilometers away from the main application locations. It is a complicated blend of chlorinated terpenoids including hundreds of distinct congeners. To distinguish between the numerous toxaphene congeners, Harun Parlar developed a nomenclature based on the gas chromatographic retention period. In terms of biodegradability, Ecotoxicity, and long-term environmental fate, Toxaphene congeners range in life expectancy from extremely long to quick degradable and biodegradable. The most frequent way of eliminating toxaphene is anaerobic degradation, however, aerobic breakdown is also possible, albeit more slowly (Guerra et al., 2017). Because of the chemical stability, environmental durability, and molecular properties of various congeners, toxaphene molecules have been biomagnified in the food chain. However, due to the natural buildup of POPs in plant biomass, there is a scarcity of research on the use of plants for phytoremediation. In this study on the absorption of toxaphene by amaranth, amaranth is employed for the first time as a bioremediation agent for toxaphene. Toxaphene was utilized on more than 100,000 acres of farms in Nicaragua, mostly in the Chinandega and Leon cotton areas.

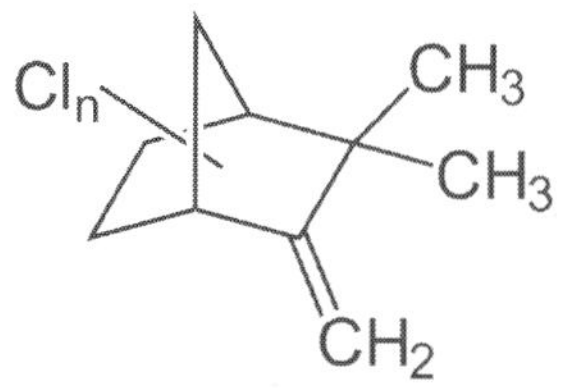

Figure 9.6: Toxaphene.

Source: By NEUROtiker - Own work, Public Domain, https://commons.wikimedia.org/w/index.php?curid=3228269

Nicaragua was one of the world's top 24 cotton producers during the cotton boom of the 1960s and 1970s. Each year, Nicaragua uses around 28-35 pesticide sprayings for cotton cultivation. Cotton farmers have discovered pesticide-resistant pests and rising pesticide use rates. Pesticides totaling 560 kg-1 have been sprayed across vast swathes of agricultural land since 1974, with yearly application rates of 31 kg ha-1. Although toxaphene was outlawed in 1993, the World Health Organization (WHO) issued a warning against its usage in 1984 due to the possibility of carcinogenic consequences. Because of its resilience, some of it lingers in agricultural soil and ultimately leaches into rivers and estuaries (García-Ruiz et al., 2017).

The overall level of toxaphene residue in Nicaragua's historic cotton areas' soils in 2002 was 44 grams per cubic meter of soil. Because of pollution, toxaphene levels in flathead grey mullet (*Mugil cephalus*) and *Anadara* spp., which are important components of indigenous cuisine, exceed the permissible range for human consumption. Toxaphene residues have been found in high concentrations in breast milk and well water. Since the advent of cotton cultivation in the 1970s, Nicaragua's best agricultural soil, the lush lowlands on the Pacific coast, has lost most of its agricultural potential.

9.8 TOXIC HEAVY METALS

Even though dangerous heavy metals are normally present in minute amounts in soil, human activity has interrupted the slow geochemical cycle of metals in many rural and urban areas, increasing the risk of health consequences for humans, plants, animals, and ecosystems. Toxic heavy metals are accumulating in the environment as a result of pollution caused by mine tailings and waste disposal; leaded gasoline and paints; sewage sludge; fertilizers; and pesticides used on land. Soil is the principal reservoir for harmful heavy metals that enter the environment. Metals must be elements to be broken down by bacteria or chemicals. This means that metals can persist in soil for a long period after they are introduced (Cogo, 2022). A variety of mechanisms influence heavy metal mobility in soil, including sorption/desorption interactions, chemical complexation with inorganic and organic ligands, and biotic and abiotic redox activities. A multitude of variables impacts heavy metal bioavailability, leaching, and toxicity in soil, including pH, sorbent type, humic and fulvic acid, root exudates, microbial metabolites, and nutrients.

Preferential flow and organic-mediated transport must be present for metals to move large distances in soil, and these conditions must be met. Because certain plants are capable of absorbing and bioaccumulating heavy metals, phytoextraction may be a low-cost strategy for remediating heavy metal-polluted soil. Chelating compounds such as EDTA (ethylene diamine tetraacetic acid), EDDS (ethylene diamine succinate), NTA (nitrilotriacetic acid), natural low molecular weight organic acids, and humic substances have been demonstrated to increase heavy metal solubility in phytoextraction. EDTA, on the other hand, is a slow-degrading complexing agent that has the potential to contaminate groundwater due to its resistance to biodegradation.

Before using chelating drugs, a risk analysis should be performed to account for these and other possible dangers.

Bioremediation Techniques That Aid inthe Construction of Sustainable Landscapes In Less Fortunate Locations

In academia, the word "marginalization" is frequently used ambiguously to describe and address a wide variety of occurrences. Residents in disadvantaged regions are usually denied full access to the opportunities, resources, and rights enjoyed by the majority of people in their country or region. Residents in rural areas of developing nations or sparsely populated areas of industrialized countries may be unable to fully participate in their country's economic, social, and political life. Many impoverished communities are unable to deal with soil pollution adequately due to a lack of economic incentives, governance, knowledge, and skills, as well as a lack of experience and resources (Frankl et al., 2016). Engineering solutions must be cost-effective, commensurate with the region's social and economic development circumstances, and provide concrete and practical benefits for bioremediation projects to be feasible in underserved areas. Because impoverished communities confront a plethora of social, ecological, and economic concerns, bioremediation programs must be adequately organized. Cotton has traditionally been farmed on the lush lowlands along Nicaragua's Pacific coast, which has some of the best agricultural soil in the country.

9.8.1 Toxic Properties of Heavy Metals

Because of human interference with nature's slow geochemical cycle of metals, many rural and urban areas have been discovered to contain significant levels of dangerous heavy metals. This has jeopardized human health, as well as the health of plants, animals, and ecosystems. Toxic heavy metals build in the environment as a result of industrial emissions and mining tailings, waste disposal, leaded gasoline and paint, leaded gasoline and paints, and land application of sewage sludge, fertilizer, and pesticide. Toxic heavy metals may be absorbed by soils, which absorb the majority of pollutants. Metals that are injected into the soil might remain there for a long time since they are not broken down by microbes or chemicals. A multitude of mechanisms impacts heavy metal mobility in soil, including sorption/desorption, chemical complexation with inorganic and organic ligands, and biotic and abiotic redox reactions. Soil parameters such as pH, sorbent type, the presence and quantity of organic and inorganic ligands such as humic acid and fulvic acid, root exudates, microbial metabolites, and nutrients all influence heavy metal bioavailability, leaching, and toxicity.

Techniques for Bioremediation That Will Assist Impoverished Areas

The term "marginalization" is used to characterize a variety of actions in various academic fields, although it is also used ambiguously in many contexts. According to the findings of this study, marginalized regions are those where individuals are routinely denied access to the same opportunities, resources, and rights as the majority of people in the country or region in question. Some countries, whether developing or developed, tend to neglect the economic, social, and political relevance of rural areas due to a lack of attention from government officials and investors. Because of a lack of economic incentives, governance, knowledge, and skills, many impoverished communities are unable to appropriately deal with soil pollution (Meli, 2013). Engineering solutions for bioremediation projects in impoverished areas must be cost-effective, compatible with the region's social and economic development level, and provide observable and verifiable benefits. Managing bioremediation programs in underserved areas is crucial since they confront a variety of social, ecological, and economic challenges.

Historically, cotton was grown in Nicaragua on the country's greatest agricultural soil, the lush lowlands of the Pacific coast, which have become less suited for cultivation over time. Toxic properties of heavy metals. Human interference with nature's slow geochemical cycle of metals has led to the discovery of harmful heavy metals in large concentrations in a variety of rural and urban habitats, posing a risk to human health, plants, animals, and ecosystems (EskandariDamaneh et al., 2022). The environment is contaminated with harmful heavy metals as a result of industrial emissions, mining tailings, trash disposal, leaded gasoline and paints, sewage sludge land application, fertilizers, and pesticides. Soils absorb the bulk of heavy metals discharged into the environment. Metals can persist in the soil for a long time after being introduced because they are not degraded by microbes or chemicals. A multitude of mechanisms impacts heavy metal mobility in soil, including sorption/desorption, chemical complexation with inorganic and organic ligands, and biotic and abiotic redox reactions.

Soil variables such as pH, sorbent type, the existence and quantity of organic and inorganic ligands such as humic and fulvic acid, root exudates, microbial metabolites, and nutrients all influence heavy metal bioavailability, leaching, and toxicity. Metals may move vast distances in the soil when conditions are appropriate, due to preferred flow and organic-facilitated transport. Because some plants absorb and bioaccumulate heavy metals, phytoextraction may be a low-cost approach for remediating heavy metal-polluted soil. Phytoextraction can be increased by increasing the solubility of heavy metals using chelating compounds such as EDTA (ethylene diamine tetraacetic acid), NTA (nitrilotriacetic acid), natural low molecular weight organic acids, and humic substances. Slow-degrading complexing agents, on the other hand, maybe a hazard due to their resistance to biodegradation and the possibility for increasing harmful metal leakage into groundwater. Before using chelating medications, a risk assessment should be undertaken to account for these possible dangers.

9.9 BIO-STIMULANT

Soil factors such as aeration, nutrient input, pH, and temperature are altered in biostimulation to stimulate the breakdown of contaminants. The act of changing the remediated environment to foster the growth of indigenous microorganisms capable of digesting pollutants is referred to as "biostimulation." Biostimulant. Improve their habitat by giving nutrients and/or electron acceptors that might otherwise be unavailable. Because of its ease of use and low cost, biostimulation may be a viable option for bioremediation in low-income areas. It is feasible to save a significant amount of time and money by exploiting soil-based microorganisms that have evolved to flourish in the local environment (Yakupoglu, Rodrigo-Comino &Cerdà, 2019). The capacity of microorganisms to obtain nutrients may be hampered by underlying geology, which may contain clays and other fine-grained minerals that restrict nutrient transfer through the soil matrix. Using organic by-products that would otherwise be discarded as a source of readily available carbon, nutrients, and bioactive compounds as a source of readily available carbon, nutrients, and bioactive compounds is one strategy for providing nutrients for bio stimulating in a way that may promote sustainable development. Organic by-products are created in large quantities all around the world. All around the world, by-products from farming and livestock, fishing, and forestry are ubiquitous. Several research articles have been conducted on the use of organic byproducts as bioremediation amendments.

As a result, molasses and bagasse are the most prevalent waste products, along with manure, blood meal, fishbone meal straw, and rice husks. The experimental topics for this doctorate dissertation were chosen from a list of specified by-products that may be employed as amendments for in situ biostimulations in undeveloped nations with tropical temperatures.

To boost aerobic PCB cleaning, biostimulation using oxygen, co-substrates, surfactants, and/or inducers, and bioaugmentation with competent PCB-degrading bacteria are utilized (Webb, Herrick, &Duniway, 2014). Biphenyl can be employed as a carbon and energy source as well as an enzymatic inducer to increase aerobic PCB co-metabolism because it is the primary and natural substrate for biphenyl dioxygenase enzymes. In numerous trials, including biostimulant and bioaugmentation, microbes were demonstrated to have minimal effect on PCB transformation, indicating that aerobic PCB-degrading bacteria are abundant in contaminated soil. The addition of O2, biphenyl, and inorganic nutrients to polluted sediments improved aerobic PCB biodegradation, however, *Alcaligenes eutrophus* H850, a PCB-degrading bacterium, did not result in any significant further transformation. The addition of sorbitantrioleate, a surfactant, to the PCB-degrading *Arthrobacter* sp. B1B and *R. eutropha* H850 when regularly applied to an Aroclor 1242-contaminated soil increased aerobic bioremediation.

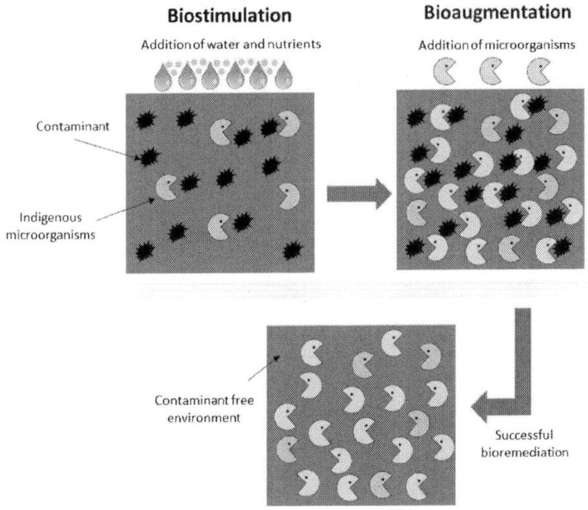

Figure 9.7: Biostimulation techniques.

Source: Sales da Silva IG, Gomes de Almeida FC, Padilha da Rocha e Silva NM, Casazza AA, Converti A, Asfora Sarubbo L. Soil Bioremediation: Over-

view of Technologies and Trends. Energies. 2020; 13(18):4664. https://doi.org/10.3390/en13184664

Despite significant study into PCB biodegradation, little is known about the biochemistry of anaerobic dechlorination and alternative aerobic degradation routes, except for the well-known biphenyl dioxygenase route. According to research, PCB breakdown can proceed through several unknown mechanisms. Among these findings is *Hydrogenophagataenio spiralis* IA3-A, which can co-metabolize lower chlorinated congeners in Aroclor 1221 at a low temperature of 5 °C. It has also recently been demonstrated that despite the lack of biphenyl dioxygenase genes, a marine *Pseudomonas* CH07 and a soil *Paenibacillus* KBC101 can digest higher chlorinated congeners.

Because Bph genes may be generated, it has been proposed to use natural inducers to accelerate PCB biodegradation in PCB-contaminated regions. When biphenyl addition is employed, additional carbon and energy sources for co-metabolism have been demonstrated to boost PCB transformation. However, due to the toxicity of biphenyl, a quest for natural inducers has begun. Plant-derived isoprenoid and flavonoid compounds have been demonstrated to be potent PCB breakdown inducers. Rhizoengineering is an exciting and closely related strategy for improving PCB biodegradation (Van Leeuwen et al., 2019). Transgenic plants or metabolic mutants can be used to modify secondary plant metabolites to promote the development of bacteria capable of digesting xenobiotics in the soil. Narasimhan found a large spectrum of phenylpropanoid compounds in plant exudates, including flavonoids, using rhizosphere metabolomics. The colonization of *Pseudomonas putida* PML2 roots was then aided by the overproduction of flavonoids in various Arabidopsis strains. *P. putida* PML2, a rhizospheric bacterium, may digest flavonoids and PCBs. When cultivated in conjunction with Arabidopsis expressing flavonoid, Pseudomonas PML2 removed 2- and 4-monochlorinated biphenyls much more than flavonoid-null mutants.

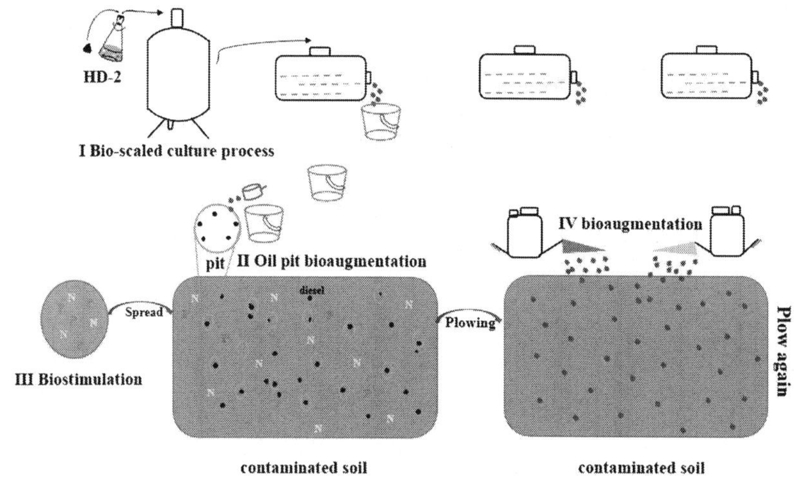

Figure 9.8: Biostimulation.

Source: Li X, Chen Y, Du X, Zheng J, Lu D, Liu Z. Ecological Response in the Integrated Process of Biostimulation and Bioaugmentation of Diesel-Contaminated Soil. Applied Sciences. 2021; 11(14):6305. https://doi.org/10.3390/app11146305

9.10 WHEY

Whey is a byproduct of milk casein removal that is used to make cheese and other dairy products. This liquid byproduct provides between 85 and 95 percent of the volume of milk, accounting for more than half of its content. Food production accounts for about half of the world's annual output of 145 million metric tons. Whatever remains is a source of contamination for the rest of the ecosystem. The chemical makeup of whey includes vitamins B1, B2, B6, and B12, as well as folic acid and lactic acid (Tsoraeva et al., 2021). Lactose can be found in whey. Although there is evidence that in situ whey application is an energy-efficient and non-disruptive approach to in situ bioremediation in low priority areas, little study has been done on the effect of whey on the biodegradation of organic contaminants. According to the MIUN study, aliphatic and aromatic hydrocarbons break down faster when mixed with milk. The whey had a good influence on biodegradation due to the easily available carbon source. The breakdown of phenanthrene, an aromatic compound, revealed a more intricate relationship between carbon sources and growth factors like B vitamins than an aliphatic molecule (hexadecane).

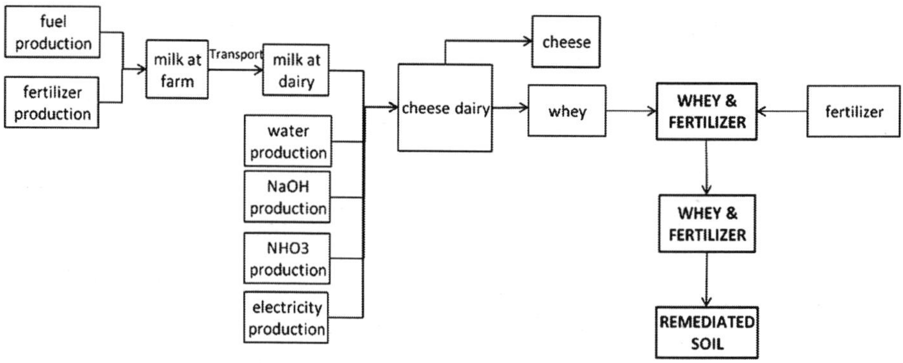

Figure 9.9: Whey in bioremediation.

Source: Jonsson, A., & Haller, H. (2014). Sustainability Aspects of In-Situ Bioremediation of Polluted Soil in Developing Countries and Remote Regions. In (Ed.), Environmental Risk Assessment of Soil Contamination. IntechOpen. https://doi.org/10.5772/57315

According to research, bioremediation of soil contaminated with diverse organic pollutants necessitates the consideration of components such as vitamins and minerals. Because of the low soil temperature and limited whey supply due to clay soil percolation limits, in situ research employing the whey method in Sweden has not given the same favorable findings as those achieved in the laboratory. Assuming the whey supply issue is resolved, the approach may be more successful in a tropical setting where mineralization rates are anticipated to be significantly higher. Because high whey concentrations delayed biodegradation while low concentrations had no impact, it is critical to select the appropriate application rates to guarantee optimal breakdown. Aside from soil type and precipitation, the optimal application rate may change according to temperature and precipitation. Repeated applications appear to accelerate the breakdown of a range of organic pollutants more than a single spray.

One of the key ingredients of this chemical is PA or pyrroligneous acid. PA is a low-value byproduct of numerous charcoal manufacturing processes that goes by a variety of names, including wood vinegar, pyrolysis oil, smoke extract, and bio-oil. At the moment, PA's only economically viable usage is as a smoke flavor in food. The Food and Drug Administration has ruled smoke flavor acceptable as long as it is used following approved manufacturing methods (Tully et al., 2015). Even though PA is typically discarded, it has been utilized in Japan for millennia to enhance agricultural yields and combat pests. Today, traditional kilns in developing nations generate the majority of the world's charcoal. Pollution of the land and air arises from improper management of gases and fumes. The pyrolysis process's condensate co-products may be collected and utilized as a fuel in industrial processes, lowering pollution. Two low-tech kilns were created to recover PA from charcoal production: the Brazilian beehive and the Casamance. Because of recent improvements in PA research and technology, charcoal manufacturing is heading in the direction of greater environmental performance without losing output or quality.

The chemical composition of PA varies depending on the feedstock source, pyrolysis temperature, and pyrolysis time. Methanol, 2-butanone, acetaldehyde, furan, furfural, acetic acid, and other volatile organic acids are commonly detected in PA. PA includes antibacterial phenolic chemicals and has been used successfully to inhibit fungal growth on wood as well as bacterial food degradation. Small dosages of PA, on the other hand, have been demonstrated to boost germination, growth, and yield in a wide variety of plants (Kadota and Niimi). Although the precise method by which PA increases development is uncertain, it has been connected to dietary components such as magnesium sulfate, calcium, magnesium, and potassium nitrates, as well as its stimulatory impact on symbiotic fungus and bacteria.

PA has been observed to enhance microbial biomass and basal respiration in heavily exploited tropical soils, indicating that PA is metabolized by microorganisms. Toxins in charcoal kiln smoke are digested by phototrophic soil bacteria, and their integration into the soil will almost certainly promote respiration. When PA was added to the growing substrate, *Lentinus edodes*, *Pholiota nameko*, and *Pleurotomum ostreatum* produced more fruit bodies (Telles, Guimarães, &Dechen, 2011). *Pseudomonas ostreatus* and *Lentinus edodes* can decompose a wide variety of hazardous compounds, including PAH and organochlorine insecticides. When PA was dosed at 3.0 percent, it was found to have the greatest influence on the development of Pleurotusostreatus. The effect diminished as concentrations increased, and no fruit bodies developed at 10%. Bacteria that can break down PA compounds are rare in most soils. It has been demonstrated that a consistent supply of these chemicals increases their prevalence in the soil near a charcoal furnace. Repeated PA dosages may increase the number of microorganisms targeted.

9.11 BIOAUGMENTATION

To improve indigenous populations, degrading microorganisms are introduced. Because native microbial communities with degrading potential are few in these areas, this technique is perfect. It also works effectively in locations where native microbial communities have been stressed as a result of recent pollution exposure (Salesa &Cerdà, 2020). Survival of seed microorganisms in a new place is difficult due to competition with native bacteria, tiny soil pores, and genetic stability. Bioaugmentation should be used only if the danger of potentially invasive species has been thoroughly examined.

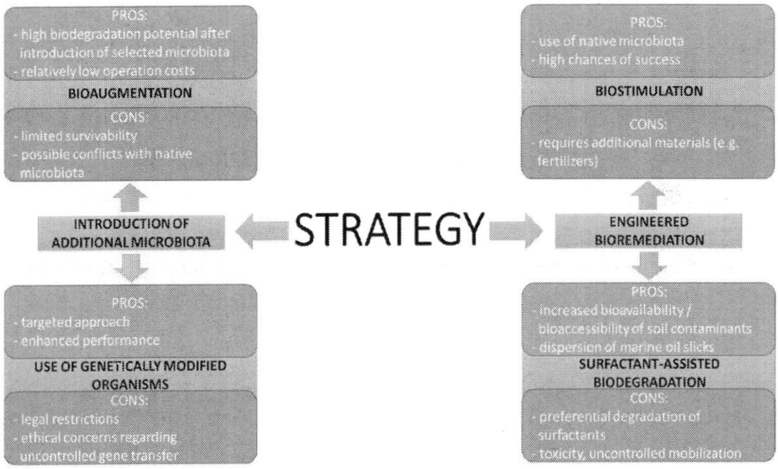

Figure 9.10: A combination of bio augmentation and bio stimulation.

Source: Ławniczak Ł, Woźniak-Karczewska M, Loibner AP, Heipieper HJ, Chrzanowski Ł. Microbial Degradation of Hydrocarbons—Basic Principles for Bioremediation: A Review. Molecules. 2020; 25(4):856. https://doi.org/10.3390/molecules25040856

Because it provides soil bacteria and nutrients, an example of compost tea as a bioaugmentation and biostimulating strategy is presented. Compost Teas (CT) are water-based soil supplements made from steeped and extracted compost. Although it is not a byproduct in and of itself, the great majority of its components are. Humic acid supplements also include molasses and fish hydrolysate, rock dust, yeast extract, and yeast extract, in addition to humic acids. It may be a realistic option for individuals in need of a cost-effective, easy-to-use, and generally embraced solution. CT was employed in agriculture by the Romans and ancient Egyptians, but it was phased out when synthetic pesticides became widely accessible, and it is now largely utilized in organic farming. Evidence implies that disease-fighting and soil-improvement abilities may be quantitatively demonstrated, although this is rarely the case. Despite its newness, a growing number of research publications published over the previous several decades have proved CT expertise's ability to prevent a wide spectrum of airborne and soil-borne plant illnesses. A variety of manufacturing factors were changed to increase plant disease resistance (such as aeration, compost supply, and nutritional supplements).

There is currently no agreement on the ideal production conditions for disease suppression, and some research has had unsatisfactory results. Aeration and nutrition supplementation are two methods for increasing the cell mass of active microorganisms. Furthermore, humic acids and yeast extracts have been shown to significantly increase the fungal population in the soil. Even though the majority of research has been conducted on NCT, aerating compost teas (ACT) to preserve aerobic conditions during the extraction process is becoming more common. In recent years, several businesses have developed that manufacture CT equipment capable of functioning in highly aerated conditions (Rodrigo-Comino, 2018). The amount of oxygen in the tea appears to alter microbial populations, necessitating a differentiation between ACT and NACT compost teas (NCT). The majority of CT research has focused on their potential for disease management in agriculture since their development. The bioremediation supplement CT, on the other hand, has been shown to boost soil microbial respiration and dehydrogenase activity considerably.

The fungal communities were dominated by Aspergillus, Penicillium, and Trichoderma, while the bacteria were dominated by *Bacillus sp., Lactobacillus sp., Micrococcus luteus, Staphylococcus sciuri, Pseudomonas putida, Burkholderia glumae, and Clavibactera gropyri*. A range of species in this genus can digest polycyclic aromatic hydrocarbons (PAHs) and organochlorine chemicals (Hu et al., 2017). The use of aeration and fertilizer has been demonstrated to considerably boost the population of the species discovered. A substantial quantity of research has been published on the microbiological makeup of ACT utilizing standard culture protocols or microscopic analyses. Based on these technologies, it is obvious that molecular methods will be required to completely comprehend the ACT's diverse bacteria.

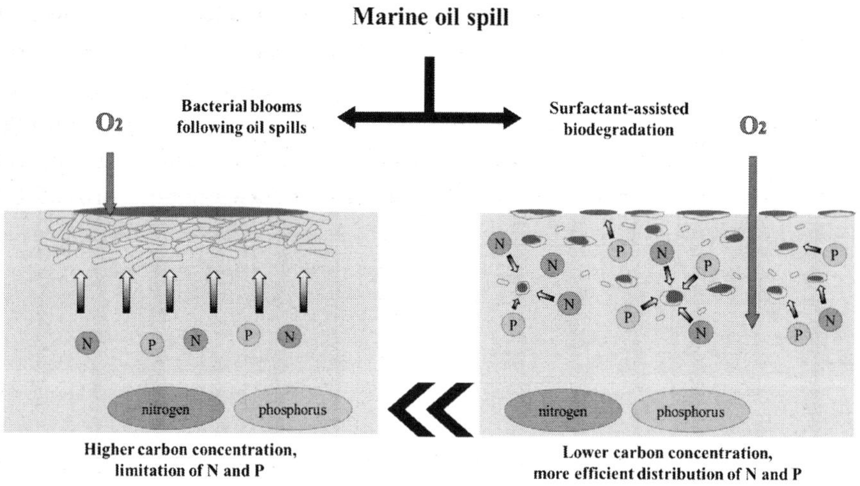

Figure 9.11: Integrated approach of bioremediation including bioaugmentation.

Source: Ławniczak Ł, Woźniak-Karczewska M, Loibner AP, Heipieper HJ, Chrzanowski Ł. Microbial Degradation of Hydrocarbons—Basic Principles for Bioremediation: A Review. Molecules. 2020; 25(4):856. https://doi.org/10.3390/molecules25040856

9.12 ECO-FRIENDLY TECHNOLOGIES

Despite problems such as the Great Recession, the Eurozone Crisis, and a slew of tragic natural disasters, the world population surged by 600 million people a decade ago! A recent example is an earthquake in Japan. Global output and industrial capacity still have a lot of room to develop, but per capita GDP growth in countries has been quite modest in recent years. The fast development of industrial production capabilities would be illogical. Keep in mind that when it comes to releasing economic and social revival, technology accelerators must be handled with caution. We have vivid memories of what occurred at Kalahandi, Orissa, India! We don't want this to happen to our lovely Earth, as depicted in Rachel Carson's book 'Silent Spring.'

As the environmental consequences of human actions become more obvious, humanity's commitment to mitigating those consequences grows greater. The notion of "Sustainable Development," which must be accompanied by economic growth, is introduced here. Previously, sustainable development was supposed to be achieved by three equal pillars:

the economy, society, and environment. A fresh strategy, on the other hand, is required. To achieve long-term growth, the economy must be viewed as a social function that flourishes in a healthy natural environment, rather than as two distinct pillars (Rodrigo-Comino et al., 2018). Any economic strategy should aim to improve house management and well-being, in other words, our surroundings. In this context, sustainable development is defined as "development that satisfies current demands while maintaining the Earth's life-support system, which is critical for the well-being of present and future generations." A healthy and vibrant planet is a prerequisite for healthy and bright existence.

If we want to have a flourishing economy, we need to invest in technology and innovation. Engineers may create and then implement excellent ideas using tools, procedures, or a working set of devices to expedite, advance, and finish otherwise complicated operations. When technology and resources are employed for the benefit of both the environment and the bottom line, both are successful.

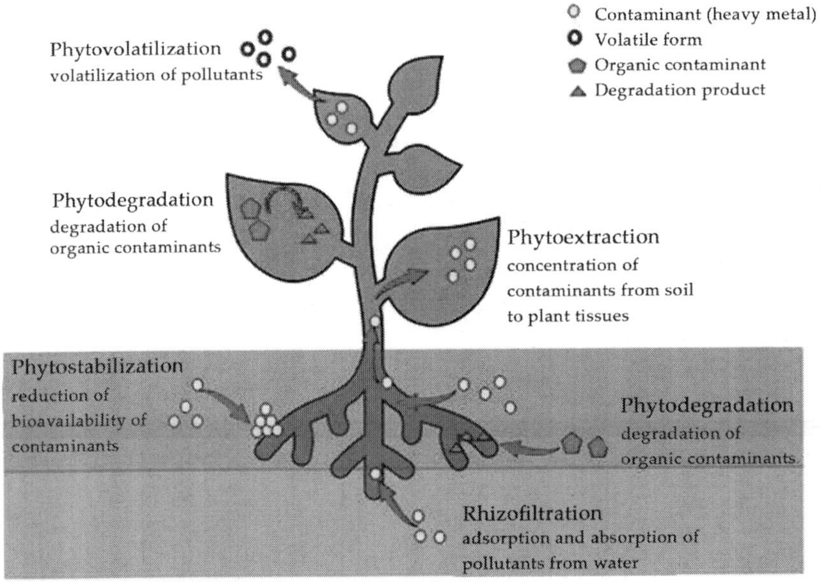

Figure 9.12: Handling soil contamination.

Source: Rigoletto M, Calza P, Gaggero E, Malandrino M, Fabbri D. Bioremediation Methods for the Recovery of Lead-Contaminated Soils: A Review. Applied Sciences. 2020; 10(10):3528. https://doi.org/10.3390/app10103528

Eco-technology encompasses not only the replacement of obsolete technology with environmentally friendly alternatives but also the efficient utilization of available resources. This benefits each country's economy and ecology. Almost every plenary meeting on the planet discusses the benefits and uses of Ecotechnology. Every year, billions of dollars are invested in its progress. Even in the industrialized world, it is regarded as an economic centre for high-end jobs, with a trend away from industry and toward services, followed by eco-technology.

Natural resource scarcity and environmental deterioration are frequently cited as indices of sustainability. Global GDP is predicted to quadruple in the next 25 years, owing mostly to growing salaries in developing markets and the adoption of comparable spending patterns in established markets. Climate change is predicted to decrease natural resources (in the broadest sense) and self-repair capabilities. To be effective, sustainability must be a bottom-up initiative.

Entrepreneurs have an important role at the bottom of the pyramid. After all, what is the ultimate purpose of the life of an entrepreneur? Income generation is less crucial than system development and avoidance of obsolescence. As a result of the foundations they lay, the level 5 businesses they leave behind will be even more successful than they were before they left.

Efficient eco-innovation implementation in emerging nations such as India and Brazil is both a foundation and a stimulant for economic progress. As a result, enterprises in these nations must capitalize on untapped market opportunities, boost productive activity, and protect market share through intellectual property rights on eco-innovation. Bio food manufacturers, renewable energy sources, and hybrid automobiles might all be differentiating factors.

Michael Porter's 1995 concept advises that corporations throughout the world take initiatives to lessen their environmental effect. Among the strategies that might be adopted are improved market access, increased profitability through product diversification, and the sale of pollution-control items. As a result, the present order must alter, whether it is transportation, water supply, sewage treatment, or enough housing. Engineers, architects, and planners may be able to assist promote a new type of urban architecture that combines climate-sensitive design with practical efficiency and a pleasant ambience (Rickson, 2014). Businesses that diversify their activities may have an edge over those that do not when it comes to planning for liability fees, fines, and litigation.

Policies at all levels of government must be modified to better match the demands of the nations and the circumstances in which they are implemented. Eco-innovation must be promoted for the economy to recover. Making policies to encourage eco-innovation, conducting impact assessments of new technologies to determine their environmental benefits and risks, identifying the infrastructures needed to assist the transition to a low-carbon, environmentally sustainable economy, and making policies to encourage necessary investments in research (both public and private) are all critical components of the eco-innovation strategy.

In the current circumstances, the efficacy of eco-innovations is a welcome challenge to the existing quo. The yearly operating expenses of conventional bioremediation systems are approximately equal to the initial investment in these systems. Because the approaches employ only natural biotic and abiotic components, they are both natural and ideal, utilizing the least amount of energy possible. As a result, they are natural. Even if it is hard to entirely replace existing technologies, eco-technological systems can be constructed and maintained alongside them to improve their performance.

The importance of environmental sustainability in our actions has become more apparent than ever before. This demonstrates that eco-technology is here to stay. India and other developing nations with a long history of peaceful cohabitation with the environment may and should be at the forefront of achieving this shift. Addressing climate change will result in tens of thousands of new enterprises, tens of millions of new jobs, and tens of billions of dollars in new income. This economy should be passed down to future generations. Nature will react and have the last laugh, endangering the lives of future generations.

However, the fact remains that no money will be distributed. It is not rare for it to be accompanied by challenges. Green technology by itself is not an economically feasible choice since it would need significant investment and would have a negative influence on the global economy's momentum. Mother Earth is trying to keep up with the rate of global change at a time when she is most in need (Poesen, 2018).

To fully account for the environmental effect, technology must be split into two phases: immediate repercussions, which are incompatible with the world's current demands, and long-term implications, which offer economic stability and environmental security. To temper the excitement for success, it is vital to understand how efforts might be made more long-term sustainable. "Be prepared to give up an immediate advantage for a long-term benefit," is the final piece of advice.

Sustainable development requires a strong human capability as well as an emphasis on eco-technology and technical innovations. As a result, long-term economic success necessitates the synergistic integration and synchronization of all three components.

When technology and the environment are combined, it is referred to as Ecotechnology. The prefix "eco" requires ecological manipulations of natural resources since they must be consistent with the long-term sustainability of natural ecosystems. The literature has several definitions and explanations, but there is no consensus on how to describe or understand Ecotechnology. According to MIUN's Department of Ecotechnology and Sustainable Building Engineering, environmentally sustainable technology is concerned with the development of future civilizations while complying with ecological restrictions. This project's main purpose was to help underdeveloped countries undertake bioremediation initiatives that go beyond pollution reduction and aid in long-term development. This presentation of the Ecotechnology concept was vital to the success of this project (Pijl et al., 2020). Adapting technology for a post-degradation and post-polluted environment is critical for the development of sustainable communities. The ultimate objective of a sustainable civilization's long-term sustainability is to maintain pollution levels that are safe for ecosystems and human health. Ecotechnology applications in the formation of long-term community relationships can help our damaged and degraded natural environment. The development of bioremediation techniques and other technologies will be required to increase the natural capacity for the recovery of ecosystems.

The full environmental assessment procedure: In the scientific literature, the phrase "integrated environmental assessment" (IEA) is used in several contexts (IEA). There is a lot of overlap in these descriptions between the phrases "integrated" and "assessment." The IEA, which is known for its mission of influencing policy and decision-making, assesses and shares environmental challenges with policymakers and decision-makers.

9.13 TECHNOLOGY

Bioremediation technology for underserved regions must be assessed and improved, therefore the notion of appropriate technology is critical to the study's purpose. The appropriate technology movement originated as a reaction to the developed world's post-World War II technical aid projects in developing nations. "Small Is Beautiful," a 1973 book by British economist E. F. Schumacher, advocates for intermediate technology, or technology that sits somewhere between the primitive and poverty-inducing tools used by many developing countries and the large, powerful technological systems used by industrialized nations. Some of the first ideas for a suitable technology were small-scale and labor-intensive, with low capital investment per worker; energy efficient; ecologically benign; owned and maintained by the local community; and low capital investment per worker.

According to several theorists like Ranis, technology may be complicated, up-to-date, capital-intensive, and so on regardless of where you are or what resources you have at your disposal. Ranis defines "suitable technology" as "the combined selection of processes and products appropriate to the greatest of society's purposes given that society's capabilities." While the number of non-governmental organizations (NGOs) committed to the promotion of suitable technology has diminished over the subsequent decades, the grassroots movement has not died.

It reached its apex in the 1970s. Many of the movement's technologies and ideas are still in use today, including solar energy, windmill-generated electricity, the use of abandoned dams for low-head hydroelectric generation, the development of biogenic methane gas for fuel, a renewed emphasis on bicycles and mass transit, recycling, and the use of natural ma. They have withstood the test of time because of their versatility (Pham et al., 2018). This initiative used suitable technology as an inspired data bank for the evaluation of technologies that address poverty, social equality, gender employment, and basic human needs in disadvantaged areas while remaining fundamentally independent from its ideological foundation.

9.14 ENVIRONMENTAL SCIENCE AND TECHNOLOGY

Ecological Engineering (EE) has been proposed as a theoretical framework for creating and mending sustainable human-ordered environments. H.T. Odom's definition of ecological engineering, widely regarded as the father of the discipline, is "environmental manipulation by a man using moderate quantities of additional energy to manage systems in which the major drivers remain natural sources." Junior and Mitch proposed a revised definition in 1989, which was later amended to "the creation of ecologically sound and healthy ecosystems that benefit both human civilization and its natural environment." It hasn't been long since others made significant contributions to the field, presenting ideas and concepts for applying ecological engineering (Zhao et al., 2015).

9.15 CONCLUSIONS

Rural regions in developing nations and sparsely inhabited areas in developed countries require a different set of solutions to polluted sites than industrial areas. According to research, using conveniently accessible organic byproducts as amendments or locally available plants for phytoremediation may be effective alternatives in economically challenged places. To be effective, bioremediation requires a detailed evaluation of the polluted site utilizing the integrated planning guide. In some scenarios where the integrated planning guide has been proved to be effective, bioremediation is being used in disadvantaged communities to meet sustainability goals beyond pollution reduction targets (Pietro & Mahajan, 2022). Despite pilot-scale research indicating that whey may considerably accelerate the biodegradation of diesel fuel in soil, the poor overall degradation rates caused by ultisol's high clay content may be a significant barrier to successful pollutant removal in full-scale applications. The vertical mobility of injected microbes was reduced in thick soils, according to the results of a soil cylinder experiment. The utilization of amaranth species capable of bioaccumulating toxaphene and other persistent organic pollutants in the region may improve phytoremediation. People and animals, on the other hand, should avoid eating the crops since the dangerous substances have been transferred to edible areas of the crops.

CHAPTER 10

SOIL PROTECTION POLICY

CONTENTS

10.1 What Is Soil Protection? 246
10.2 European Union (EU) Soil Policy 248
10.3 Existing Laws on Soil Protection 250
10.4 The European Green Deal 252
10.5 The Resource Conservation and Recovery Act (RCRA) 254
10.6 Assessment of Current Soil Protection Strategies 255
10.7 RCA Reports Usefulness Can Be Improved 256
10.8 Longstanding Social Perspectives and Customs 257

10.1 WHAT IS SOIL PROTECTION?

Soil protection fundamentally includes protecting various soil functions such as: organic soil functions that serve as habitats; separating and buffering; organic matter disintegration and decomposition. These roles are closely associated with soil use and applications, especially soil used for agriculture. Without soil protection, erosion would become rampant. Soil is crucial for food production. Crops require it to grow, and livestock needs foliage for feed. Soil conservation can assist in combating food insecurity and creating healthy communities (Priori et al., 2020).

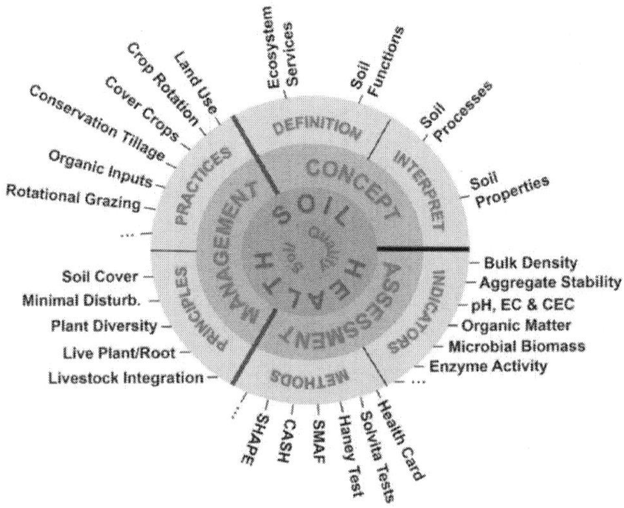

Figure 10.1: Healthy soils create benefits for people, nature and the climate.

Source: Guo M. Soil Health Assessment and Management: Recent Development in Science and Practices. Soil Systems. 2021; 5(4):61. https://doi.org/10.3390/soilsystems5040061

Environmental Legislations

1. Green chemistry

The concept of green chemistry applies innovative scientific solutions to environmental problems encountered in laboratories. The Twelve Principles of Green Chemistry were developed in 1991 by organic chemists Paul T. Anastas and John C. Warner working in EPA's Office of Pollution Prevention and Toxins. These principles can be broken down into "reduce risk" and "minimize environmental footprint".

The goal of green chemistry is to reduce chemical-related human health impacts and fundamentally eliminate environmental pollution through targeted sustainable prevention programs (Van Straalen, 2002). Green chemistry is looking for an alternative, environmentally friendly reaction media while striving to increase reaction rates and reduce reaction temperatures.

2. Transport and management of toxic substances

In China, pursuant to Article 20 of the Toxic Chemical Substances Management Law of 1991, this Regulation is formulated to regulate the transportation of toxic chemical products publicly disclosed by sea or by land in accordance with this Law, in the quantities specified in Article 3. Before the transportation of toxic chemical substances, relevant documents must be presented to the competent authority of the place of origin of transportation (Kowalska et al., 2018). The documents for submission include: 1. A six-part transport note; a copy of the license, registration certificate, or other document authorizing the handling of the toxic chemical approved by a competent authority; The transportation plan stipulated in Article 84 of the Road Traffic Act.

3. Waste management

In the UK, Section 34 of the Environmental Protection Act 1990 imposes a duty of care on those involved in waste disposal. This duty of care applies to anyone who produces, imports, transports, stores, handles or disposes of controlled waste, or who acts as a broker in the control of that waste. From 1 January 2015, new UK waste regulations will require companies to separate recyclable materials (paper, plastic and glass) from other waste. The amendment to the EU Waste Framework Directive aims to ensure improved quality and quantity of recycling across the UK.

4. Cleanup of contaminated soils

EU Directive 2004/35/EG creates a comprehensive liability regime for environmental damage. The Directive applies the polluter pays principle, where polluters are responsible for environmental damage. The responsible party is the "operator" who engages in some of the hazardous activities listed in this policy, and the operator who engages in the risk or potential risk activities listed in this policy shall bear full responsibility (no fault) for the environmental damage caused by it (Kowalska et al., 2018).

5. Management of radioactive wastes/mixed wastes

Radioactive waste in Japan is divided into two categories based on its activity, namely low-level radioactive waste (LLW) and high-level radioactive waste (HLW). The upper limit on the concentration of radionuclides allowed in waste disposed of from a reactor plant. These caps are based on reports issued by Japan's Nuclear Safety Commission (NSC) and used to prepare license applications (Shaner, 2017). Japan has been processing LLW from nuclear reactors since 1992. Regarding the disposal of high-level vitrified radioactive waste resulting from the reprocessing of spent fuel, the Act on Final Disposal of Designated Radioactive Waste came into force in June 2000, providing information on the nature of the implementing agencies involved in the selection process and the means of accumulating the necessary funds for implementation.

6. Oil pollution prevention legislation, pesticide legislation

The Oil Pollution Act (OPA) of 1990 simplified and enhanced EPA's ability to avoid and respond to disastrous oil spills. Congress passed the Oil Pollution Act of 1990 (OPA 90) in part to minimize the occurrence of oil spills through preventive strategies and to reduce the impact of future oil spills by enhancing preparedness (Rao, 2014). Section 4115 of the act requires tankers operating in U.S. waters to have double hulls. OPA also requires storage facilities and ships to submit plans to the federal government detailing how they will deal with the massive emissions.

10.2 EUROPEAN UNION (EU) SOIL POLICY

10.2.1 2030 EU Soil Plan

The 2030 EU soil plan establishes a system and practical methods to protect and regenerate soils, and guarantee that they are sensibly used. It establishes a goal and targets to accomplish healthy soils by 2050, with substantial action plans by 2030. The EU soil plan for 2030 replaces the previous 2006 Thematic Stratagem for Soil Protection.

10.2.2 The EU Soil Thematic Plan

The Thematic Plan for Soil Protection distinguishes the primary soil risks in the EU including landslides, erosion, floods, reduced soil organic content, salinization, pollution, compaction, fixing, and reduced soil biodiversity. It comprises the Communication (COM (2006) 231) from the Commission to different European Institutions, a proposition for Impact Assessment (SEC (2006) 620) and Directive (COM (2006) 232).

In 2012, the European Commission distributed a policy report on the execution of the Strategy and progressive actions (COM (2012)046). The report gives an outline of the activities done by the European Commission to carry out the four plans of the Plan, to be specific awareness raising, research, incorporation, and legislation. It additionally presents soil degradation patterns both in Europe and internationally, including limitations to guarantee protection.

After the withdrawal of the official proposition because of the resistance of smaller nations in the Council, the Commission established an Expert Group in 2015 ordered by the Member States to consider how soil quality matters could be addressed through a designated and proportionate risk-oriented approach inside a binding legal structure (Pimentel & Burgess, 2013).

Considering the multi-sectoral nature of soil matters and the variety of ecological and socio-economic tensions and governance factors across Europe, various EU policy tools and Member State levels exist that either unequivocally reference soil risks or soil functions; or certainly provide some kind of soil protection. Notwithstanding, a report examining existing soil protection approaches and measures, and distinguishing key gaps in soil protection, mentioned that EU level policy tools lack a logical and strategic policy structure for safeguarding Europe's soils satisfactorily.

10.3 EXISTING LAWS ON SOIL PROTECTION

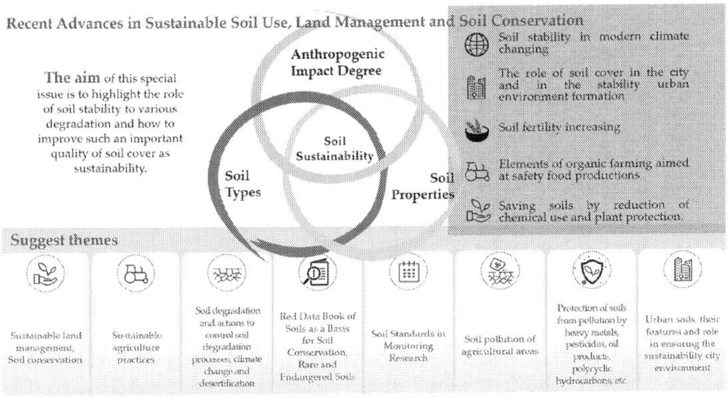

Figure 10.2: Soil conservation is essential in ensuring food security.

Source: https://www.mdpi.com/journal/sustainability/special_issues/sustainable_soil_use

At the EU level, there is no all-encompassing structure that decisively defines policy requirements or soil protection parameters. Soil protection results in different jurisdictions are for the most part determined as an outcome of assigning environmental objectives which are not expressly soil based, like lessening pollution, counterbalancing GHG emissions, and avoiding other ecological risks.

- Environmental Liability Policy

Policy 2004/35/EC on environmental obligation concerning the avoidance and restoration of environmental destruction (ELD) lays out a structure depending on the polluter pays policy to forestall and remedy environmental destruction. Other than a typical structure on damage remediation to water and natural environments, it additionally establishes the proper procedures to remediate land destruction (for example, to guarantee that pollutants are managed such that the damaged land no longer presents any major risk of negatively impacting human health).

- Industrial Emissions Policy

Policy 2010/75/EU on industrial discharges (the Industrial Emissions Directive or IED) is the primary EU tool for managing pollutant outflows from industrial establishments. It seeks to accomplish a higher degree of protection of human health, including the environment holistically by diminishing unsafe industrial discharges across the EU (Nearing, Lane, &

Lopes, 2017). It offers an incorporated approach to prevent and manage emissions into soil, air, and water for waste management, and also guarantees that the installation process doesn't cause degeneration of the soil and subterranean water.

- Environmental Impact Assessment Policy

The EIA Policy (85/337/EEC) has been operational since 1985 and has been revised multiple times, in 1997, 2003 and 2009. This Policy will apply to the appraisal of the environmental impacts of public and private programs which are probably going to affect the environment. The ecological impact evaluation will distinguish, define and review in a suitable way the immediate and indirect impacts of an activity on: people, fauna and vegetation, air, soil, water, environment and the landscape, physical resources, and so forth

- Sewage Sludge Policy

The Sewage Sludge Policy (86/278/EEC) attempts to promote the utilization of sewage waste in agriculture and to manage its application so as to avoid destructive impacts on soil, flora, fauna and humans. Sewage sludge use should not hinder soil quality and of farm products. To this goal, it restricts the use of raw sewage on agrarian land except if it is infused or fused with the soil. Treated sewage is characterized as having gone through "organic, chemical or heat treatment, extensive storage or some other suitable process so as fundamentally to decrease its fermentability and the health risks emanating from its usage".

- Policy on Fertilizers

Guideline (EU) 2019/1009 describes the meaning of 'EU fertlising products' and provides rules on making them market accessible. Among others, it additionally describes limits for impurities present in fertilizers, especially Cadmium, to reduce soil contamination.

- Mercury Regulation

Guideline (EU) 2017/852 covers the entire life stages of mercury. It provides measures and conditions around the utilization, storage and trade of mercury, its mixes and combinations, the production, usage, and exchange, mercury-added products, and mercury waste management. The directive seeks to guarantee a high degree of protection of human wellness, and safeguarding the environment from anthropogenic discharges and mercury emissions and its mixtures.

- Land Use, Changes and Forestry Policy

Guideline (EU) 2018/841 on the incorporation of ozone-depleting emissions and land use removals, land use changes and forestry service in the 2030 environment and energy structure, establishes a binding commitment for every Member State to guarantee that outflows from land use are totally remunerated by an identical expulsion of CO_2 from the climate through effective action plan (Nigussie et al., 2015).

- Common Agricultural Policy

The Common Agriculture Policy (CAP), and four related guidelines, establish the various components of the CAP framework (for example rules for instant payments to farmers under the EU guideline 1307/2013; management, financing, and assessment of the common agricultural approach EU guideline 1306/2013; and so forth. The CAP is a significant economic driver for farming choices across the EU and can foster soil protection in both farming and forestry services through Member States, including land administrators' execution of its actions and related commitments. Soil is among the fundamental resources for farming and forestry production. The CAP goal of sustainable management of natural assets and environmental action, and more particularly the delivery of natural public products and the quest for environmental change mitigation and adaptation, are certainly significant to soil protection and upgrading.

10.4 THE EUROPEAN GREEN DEAL

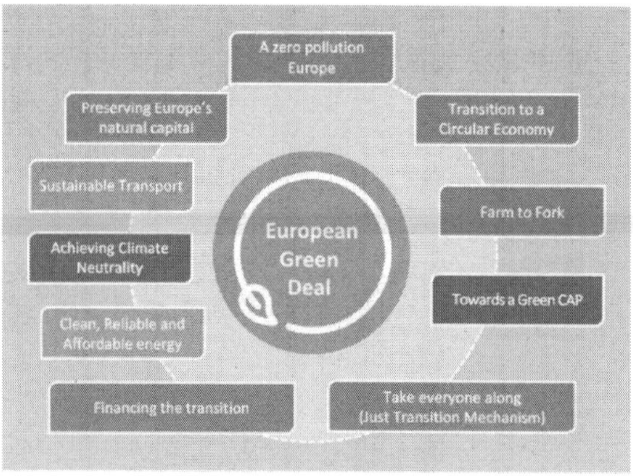

Figure 10.3: The European Green Deal will ensure clean, and sustainable energy through proper soil management.

Source: https://www.mdpi.com/about/announcements/2233

In Dec 2019, the European Commission introduced the European Green Deal, which reprograms the Commission's obligation to handling the environment and nature linked complications. The European Green Deal is a reaction to these difficulties through a fresh development procedure that intends to change the EU into an equal and prosperous society, with a cutting edge, resource efficient and competitive economy where there are no net outflows of ozone-harming substances in 2050 and where economic development is decoupled from resource use. It likewise seeks to protect, preserve and improve the EU's capital, and safeguard the health and prosperity of residents from climate-related risks (Molla&Sisheber, 2017).

In accordance with the EU Green Deal, the Commission approved in May 2020 the fresh EU Biodiversity Plan for 2030: Bringing nature back to life - a complete and ambitious extensive policy for safeguarding nature and stopping the deterioration of environments.

Fundamental components for land and soil use in the EU Biodiversity Plan for 2030 are listed below:

- A fresh EU Nature Restoration Policy will assist with working on the strength of existing and new protected regions, and provide versatile nature to all environments. This implies decreasing pressure on living spaces and species, and guaranteeing practical utilization of environments. It likewise implies supporting the reclamation of nature, limiting soil fixing and expanding suburbia, and handling pollution and intrusive alien species.
- The Commission will advance a proposition for legally controlling EU nature rebuilding targets in 2021 to reestablish degraded environments, specifically those with the most capacity to catch and store carbon and to forestall and lessen the effect of catastrophic events. The impact appraisal will likewise examine the probability of an EU-wide approach to plan, evaluate and accomplish suitable environmental conditions, so they can provide benefits like environmental regulation, soil health, water regulation, fertilization and disaster avoidance and protection (Moges& Holden, 2007).
- It is crucial to increase efforts towards ensuring soil fertility, and decrease soil erosion and nutrient over-utilization, while improving soil organic content levels. This can be achieved by taking on a reasonable soil management plan.

- Critical advancement is additionally required to distinguish contaminated soil locations, regenerate degraded soils, distinguish the conditions for their great environmental status, offer reclamation goals, and work on the observation of soil quality.
- To resolve these issues in a far-reaching way and assist with satisfying EU and global responsibilities on land-degradation neutrality, the Commission updated the EU Soil Thematic Policy in 2021.
- The Zero Pollution Action Proposal for Soil, Air, and Water that the Commission undertook in 2021, will review a scope of chemicals, including pesticides.
- Soil fixing and restoration of degraded brownfields will be discussed in the forthcoming Strategy for a Circular Economy Action Plan, and Sustainable Built Environment.

10.5 THE RESOURCE CONSERVATION AND RECOVERY ACT (RCRA)

The Resource Conservation and Recovery Act (RCRA) is a United States federal law that regulates the handling of hazardous and toxic materials. The main purpose of this act is to protect human health, the environment, and the public from potential pollution caused by hazardous waste (Meli, 2013).

The Resource Conservation and Recovery Act (RCRA) is a federal law passed in 1976. It is essentially the Toxic Substances Control Act (TSCA) with amendments and additions. RCRA governs the disposal of hazardous and toxic wastes, and establishes specific requirements for the proper treatment, storage, and disposal of these materials. RCRA has a dramatic impact on the environment. By regulating how hazardous and toxic materials are handled and disposed of, RCRA has resulted in a lower environmental impact and a cleaner environment (Zhao et al., 2015).

In addition, RCRA has helped to create a more sustainable society by promoting recycling and the proper disposal of products that would otherwise be disposed of as waste. RCRA also impacts businesses and individuals. By establishing specific requirements for the treatment, storage, and disposal of hazardous and toxic materials, RCRA has helped to save businesses money and time. In addition, RCRA has led to the development of efficient and secure waste handling and storage systems. RCRA is a key law governing the management of hazardous and toxic wastes(Hu et al., 2017). By regulating how these materials are handled and disposed of, RCRA has helped to create a cleaner environment and a more sustainable society.

Figure 10.4: The USDA pays farmers to improve their soils.

Source: https://civileats.com/2021/01/12/why-arent-usda-conservation-programs-paying-farmers-more-to-improve-their-soil/

10.6 ASSESSMENT OF CURRENT SOIL PROTECTION STRATEGIES

In conducting RCA work, USDA gathered an immense measure of data - a large number of which had previously been created - but didn't utilize this information to evaluate every one of its present projects. USDA's program administrator mentioned that the projects were not assessed in light of the fact that (1) the time periods provided were excessively severe and (2) individual organizations inside USDA had diverse assessment criteria - from no rules to measures underscoring subsidizing and individuals employed. Moreover, the draft reports were everything USDA could manage in the time allowed. However, adequate time would have been accessible to assess each program in case the USDA had not required a year to perceive the full extent of the RCA mandate. Field work forces had issues creating data for RCA reports USDA wanted to assemble the public's opinions on what significant soil and water issues and concerns challenge the United States by creating questionnaires. However, the surveys created by USDA central command were not shipped off USDA State workplaces in the time anticipated (Ma, Zhao, & Zhu, 2021). Subsequently, in some U.S. States, staff created their own soil protection questionnaires in view of the data they expected to be recalled for the public questionnaires. After the general questionnaires were shipped off the States, regional preservationists in the different States with their individual questionnaires experienced issues deciphering and rendering the data from State questionnaires to the USDA field authorities.

The RCA program director mentioned that a portion of the autonomously developed State analysis results couldn't be utilized in RCA reports since these polls served just personal interest groups. As a component of their field systems, USDA authorities held local public meetings for circulating the questionnaire and inform the general public and request its perspectives on soil and water protection. In the state of Idaho, for example, local USDA authorities mentioned that meetings were conducted at inconvenient times, accordingly hindering holistic public involvement. For example, a few meetings were conducted during the harvest period, which affected participation and finishing up questionnaires. In Nebraska, participation sheets indicate that a few groups were prevalently comprised of USDA staff. Moreover, just 3,004 of the 30,000 State questionnaires delivered were returned.

10.7 RCA REPORTS USEFULNESS CAN BE IMPROVED

Since the reports will be utilized as decision-making instruments by both the U.S. Congress and USDA for imminent water and soil resource management, the report information needed by the act should be precise r and extra information that isn't needed by the policy must be incorporated. To make savvy program choices and practically project soil and water savings, USDA must develop and measure the many impacts of water protection, including diminished energy needs, improved soil and water quality, and much more. The institutional and social requirements that will improve soil and water preservation methods must be considered, consequently determining how soil and water resources can be saved (Lim et al., 2020).

These limitations incorporate State soil and water rights, legislations, and practices and perspectives toward water usage. Implementing water protection systems would require reduced energy since the quantity of water pumped to grow crops would be diminished. While some studies have been done to determine how much energy would be saved, significant energy reserve funds would obviously come from improved soil and water preservation. Notwithstanding, USDA has not defined what impact the extra energy savings shall have on soil and water preservation.

10.8 LONGSTANDING SOCIAL PERSPECTIVES AND CUSTOMS

Longstanding social perspectives and customs on soil and water development and use are considered by different specialists as significant limitations to executing water preservation practices. Long-standing models of water utility that are administered by customs and habits are not readily changed. For instance, farmers are regularly hesitant to introduce improved watering systems since they differ from recognizable irrigation systems and can be costly. However, older irrigation systems may not be as efficient in soil and water management.

Figure 10.5: Conservation programs prevent erosion and conserve water quality.

Source: https://www.canr.msu.edu/news/conservation_programs_help_reduce_rosion_and_protect_water_quality

Generally, the Soil and Water Resources Conservation Act (RCA) planned that USDA assesses, on a continuous basis, every one of its 34 existing soil and water preservation programs and occasionally report the outcomes to the Congress. Nevertheless, the USDA's 1980 final reports did not contain an assessment of each soil and water protection program. This inability to conform to the act's purpose is expected principally to USDA's initial choice to restrict the examination to SCS programs. After a year, when the audit was extended to incorporate any remaining USDA programs, a brief period was left to follow through with the demonstration (Li et al., 2018). In case USDA had begun the full assessment right away, it should have had the option to actually assess every one of the 34 soil and water protection programs. USDA would likewise improve the practicality of its reports by including extra appropriate information. This information must incorporate the impacts that preservation benefits and limitations and institutional and attitudinal requirements have on sustainable soil and water preservation. This data would be important for decision-makers in planning projects to fulfill future soil and water conservation requirements.

REFERENCES

1. Beilicci, E., Beilicci, R, &Stefănescu, C. (2014). Aspects of optimization of soil erosion control systems. *Research Journal of Agricultural Science*, *46*(1).
2. Blake, W. H., Kelly, C., Wynants, M., Patrick, A., Lewin, S., Lawson, J., &Ndakidemi, P. (2021). Integrating land-water-people connectivity concepts across disciplines for co-design of soil erosion solutions. *Land Degradation & Development*, *32*(12), 3415-3430.
3. Blake, W. H., Rabinovich, A., Wynants, M., Kelly, C., Nasseri, M., Ngondya, I., &Ndakidemi, P. (2018). Soil erosion in East Africa: an interdisciplinary approach to realising pastoral land management change. *Environmental Research Letters*, *13*(12), 124014.
4. Chuchkalov, S., Fadeev, I., & Alekseev, V. (2020). Effect of synthetic detergents on soil erosion resistance. *KnE Life Sciences*, 489-496.
5. Cerda, A., Rodrigo-Comino, J., Novara, A., Brevik, E. C., Vaezi, A. R., Pulido, M.,&Keesstra, S. D. (2018). Long-term impact of rainfed agricultural land abandonment on soil erosion in the Western Mediterranean basin. *Progress in Physical Geography: Earth and Environment*, *42*(2), 202-219.
6. Cogo, N. P. (2022, February). Soybean Production and the Soil Erosion Problems and Solutions–South America. In *World Soybean Research Conference III: Proceedings* (pp. 1158-1165). CRC Press.
7. Ettler, V. (2016). Soil contamination near non-ferrous metal smelters: A review. *Applied geochemistry*, *64*, 56-74.

8. EskandariDamaneh, H., Khosravi, H., Habashi, K., EskandariDamaneh, H., &Tiefenbacher, J. P. (2022). The impact of land use and land cover changes on soil erosion in western Iran. *Natural Hazards, 110*(3), 2185-2205.
9. Fenta, A. A., Tsunekawa, A., Haregeweyn, N., Tsubo, M., Yasuda, H., Kawai, T., ...& Sultan, D. (2021). Agroecology-based soil erosion assessment for better conservation planning in Ethiopian river basins. *Environmental Research, 195*, 110786.
10. Frankl, A., Deckers, J., Moulaert, L., Van Damme, A., Haile, M., Poesen, J., & Nyssen, J. (2016). Integrated solutions for combating gully erosion in areas prone to soil piping: innovations from the drylands of Northern Ethiopia. *Land Degradation & Development*, *27*(8), 1797-1804.
11. García-Ruiz, J. M., Beguería, S., Lana-Renault, N., Nadal-Romero, E., &Cerdà, A. (2017). Ongoing and emerging questions in water erosion studies. *Land Degradation & Development*, *28*(1), 5-21.
12. Guerra, A. J. T., Fullen, M. A., Jorge, M. D. C. O., Bezerra, J. F. R., &Shokr, M. S. (2017). Slope processes, mass movement and soil erosion: a review. *Pedosphere*, *27*(1), 27-41.
13. Hu, B., Jia, X., Hu, J., Xu, D., Xia, F., & Li, Y. (2017). Assessment of heavy metal pollution and health risks in the soil-plant-human system in the Yangtze River Delta, China. *International journal of environmental research and public health*, *14*(9), 1042.
14. Ha, N. M., Van Dung, N., & Ngoc, H. H. (2013). Application of USLE and GIS tool to predict soil erosion potential and proposal land cover solutions to reduce soil loss in Tay Nguyen. *Vietnam Journal of Earth Sciences*, *35*(4), 403-410.
15. Haregeweyn, N., Tsunekawa, A., Nyssen, J., Poesen, J., Tsubo, M., TsegayeMeshesha, D., ...&Tegegne, F. (2015). Soil erosion and conservation in Ethiopia: a review. *Progress in Physical Geography*, *39*(6), 750-774.
16. Ighodaro, I. D., Lategan, F. S., & Yusuf, S. F. (2013). The impact of soil erosion on agricultural potential and performance of Sheshegu community farmers in the Eastern Cape of South Africa. *Journal of Agricultural Science*, *5*(5), 140.
17. Issaka, S., &Ashraf, M. A. (2017). Impact of soil erosion and degradation on water quality: a review. *Geology, Ecology, and Landscapes*, *1*(1), 1-11.

18. Jomaa, S., Barry, D. A., Heng, B. C. P., Brovelli, A., Sander, G. C., &Parlange, J. Y. (2013). Effect of antecedent conditions and fixed rock fragment coverage on soil erosion dynamics through multiple rainfall events. *Journal of Hydrology*, *484*, 115-127.
19. Kavian, A., Alipour, A., Soleimani, K., Gholami, L., Smith, P., & Rodrigo-Comino, J. (2019). The increase of rainfall erosivity and initial soil erosion processes due to rainfall acidification. *Hydrological Processes*, *33*(2), 261-270.
20. Kumarasinghe, U. (2021). A review on new technologies in soil erosion management. *J. Res. Technol. Eng*, *2*, 120-127.
21. Kheyrodin, H. A. M. I. D. (2016). Modeling soil erosion in Iran. *Innovat Intl. J. Med. Pharmac. Sci*, *1*(1), 1-12.
22. Kelly, C., Wynants, M., Munishi, L. K., Nasseri, M., Patrick, A., Mtei, K. M., &Ndakidemi, P. A. (2020). 'Mind the Gap': Reconnecting Local Actions and Multi-Level Policies to Bridge the Governance Gap. An Example of Soil Erosion Action from East Africa. *Land*, *9*(10), 352.
23. Keesstra, S., Nunes, J., Novara, A., Finger, D., Avelar, D., Kalantari, Z., &Cerdà, A. (2018). The superior effect of nature based solutions in land management for enhancing ecosystem services. *Science of the Total Environment*, *610*, 997-1009.
24. Kowalska, J. B., Mazurek, R., Gąsiorek, M., &Zaleski, T. (2018). Pollution indices as useful tools for the comprehensive evaluation of the degree of soil contamination–A review. *Environmental geochemistry and health*, *40*(6), 2395-2420.
25. Lefèvre, C., Balks, M., Nziguheba, G., &Poch, R. M. (2021). Guest editorial—Policies and practices in action to address soil erosion: A special issue from the Global Symposium on Soil Erosion 2019. *Land Degradation & Development*, *32*(12), 3389-3392.
26. Lal, R. (2019). Accelerated soil erosion as a source of atmospheric CO2. *Soil and Tillage Research*, *188*, 35-40.
27. Li, S., Li, Y., Huang, X., Hu, F., Liu, X., & Li, H. (2018). Phosphate fertilizer enhancing soil erosion: effects and mechanisms in a variably charged soil. *Journal of soils and sediments*, *18*(3), 863-873.
28. Lim, Y. H., Kim, S. W., Nam, S., Chun, K. W., & Kim, M. (2020). A Comparison of Current Trends in Soil Erosion Research Using Keyword Co-occurrence Analysis. *Korean Journal of Environment and Ecology*, *34*(5), 413-424.

29. Ma, X., Zhao, C., & Zhu, J. (2021). Aggravated risk of soil erosion with global warming–A global meta-analysis. *Catena, 200*, 105129.
30. Meli, M. (2013). *Effects of Spatial Heterogeneity in Soil Contamination: An Ecological Modelling Approach.* Department of Environmental, Social and Spatial Change, Roskilde University, 68-112.
31. Meshesha, D. T., Tsunekawa, A., Tsubo, M., &Haregeweyn, N. (2012). Dynamics and hotspots of soil erosion and management scenarios of the Central Rift Valley of Ethiopia. *International Journal of Sediment Research*, *27*(1), 84-99.
32. Mekuria, W., Veldkamp, E., Haile, M., Gebrehiwot, K., Muys, B., & Nyssen, J. (2009). Effectiveness of exclosures to control soil erosion and local communities perception on soil erosion. *African Journal of Agricultural Research*, *4*(4), 365-377.
33. Moges, A., & Holden, N. M. (2007). Farmers' perceptions of soil erosion and soil fertility loss in Southern Ethiopia. *Land Degradation & Development*, *18*(5), 543-554.
34. Molla, T., &Sisheber, B. (2017). Estimating soil erosion risk and evaluating erosion control measures for soil conservation planning at Koga watershed in the highlands of Ethiopia. *Solid Earth*, *8*(1), 13-25.
35. Nigussie, H., Tsunekawa, A., Jan, N., Jean, P., Tsubo, M., Derege, T. M., &Firew, T. (2015). Soil erosion and conservation in Ethiopia: a review. *Progress in Physical Geography*, *39*(6), 750-774.
36. Nearing, M. A., Lane, L. J., & Lopes, V. L. (2017). Modeling soil erosion. In *Soil erosion research methods* (pp. 127-158). Routledge.
37. Pimentel, D., & Burgess, M. (2013). Soil erosion threatens food production. *Agriculture*, *3*(3), 443-463.
38. Priori, S., Pellegrini, S., Vignozzi, N., &Costantini, E. A. (2020). Soil Physical-Hydrological Degradation in the Root-Zone of Tree Crops: Problems and Solutions. *Agronomy*, *11*(1), 68.
39. Pietro, P. D., & Mahajan, R. R. (2022). Erosion Control Solutions with Case Studies. In *Scour-and Erosion-Related Issues* (pp. 71-94). Springer, Singapore.
40. Pham, T. G., Degener, J., &Kappas, M. (2018). Integrated universal soil loss equation (USLE) and Geographical Information System (GIS) for soil erosion estimation in A Sap basin: Central Vietnam. *International Soil and Water Conservation Research*, *6*(2), 99-110.

41. Pijl, A., Reuter, L. E., Quarella, E., Vogel, T. A., &Tarolli, P. (2020). GIS-based soil erosion modelling under various steep-slope vineyard practices. *Catena*, *193*, 104604.
42. Poesen, J. (2018). Soil erosion in the Anthropocene: Research needs. *Earth Surface Processes and Landforms*, *43*(1), 64-84.
43. Rao, L. N. (2014). Environmental impact of uncontrolled disposal of e-wastes. *International Journal of ChemTech Research*, *6*(2), 1343-1353.
44. Rickson, R. J. (2014). Can control of soil erosion mitigate water pollution by sediments?. *Science of the Total Environment*, *468*, 1187-1197.
45. Rodrigo-Comino, J., Davis, J., Keesstra, S. D., &Cerdà, A. (2018). Updated measurements in vineyards improves accuracy of soil erosion rates. *Agronomy Journal*, *110*(1), 411-417.
46. Rodrigo-Comino, J. (2018). Five decades of soil erosion research in "terroir". The State-of-the-Art. *Earth-Science Reviews*, *179*, 436-447.
47. Shaner, D. L. (2017). Physiological effects of the imidazolinone herbicides. In *The Imidazoline Herbicides* (pp. 129-138). CRC Press.
48. Telles, T. S., Guimarães, M. D. F., & Dechen, S. C. F. (2011). The costs of soil erosion. *RevistaBrasileira de Ciência do Solo*, *35*(2), 287-298.
49. Tully, K., Sullivan, C., Weil, R., & Sanchez, P. (2015). The state of soil degradation in Sub-Saharan Africa: Baselines, trajectories, and solutions. *Sustainability*, *7*(6), 6523-6552.
50. Tsoraeva, E., Alborova, P., Bazaeva, L., Khanaeva, D., & Kozyrev, B. (2021). Modern innovative and unconventional methods to combat soil erosion. In *E3S Web of Conferences* (Vol. 284, p. 02003). EDP Sciences.
51. Van Leeuwen, C. C., Cammeraat, E. L., de Vente, J., &Boix-Fayos, C. (2019). The evolution of soil conservation policies targeting land abandonment and soil erosion in Spain: A review. *Land use policy*, *83*, 174-186.
52. Webb, N. P., Herrick, J. E., &Duniway, M. C. (2014). Ecological site-based assessments of wind and water erosion: Informing accelerated soil erosion management in rangelands. *Ecological Applications*, *24*(6), 1405-1420.

53. Yakupoglu, T., Rodrigo-Comino, J., &Cerdà, A. (2019). Potential benefits of polymers in soil erosion control for agronomical plans: A laboratory experiment. *Agronomy*, *9*(6), 276.

54. Zhang, W., & Huang, B. (2015). Soil erosion evaluation in a rapidly urbanizing city (Shenzhen, China) and implementation of spatial land-use optimization. *Environmental Science and Pollution Research*, *22*(6), 4475-4490.

INDEX

A

Accumulation index 81
acidification 78, 92
Acid mine drainage (AMD) 171
acid rain 198
agrochemicals 108, 136
alcohols 21
alkylbenzenes 224
Alzheimer's disease 149
animal breeding 78, 79

B

benzopyrene 108
bioaccumulation 35, 36
biodiversity 2
biomass 114
Bioremediation 217, 218, 219, 227, 228, 230, 233, 235, 237, 238, 241

C

carbonates 21
carbon dioxide 217, 218, 222
Central Nervous System (CNS) 34
chemical pollution 78
chronic diseases 150
Clay soils 222
Common Agriculture Policy (CAP) 252
computational modeling 108
copper contamination 152
Critical advancement 254
Crushing 171
cultivation 216, 217, 225, 226, 228

D

deforestation 36
diarrhea 149
diesel range organics (DRO) 224

E

earth's crust 2
earth's surface 2, 15, 17
Earthworms 223
ecological equilibrium 35
ecological risk index 82
ecosystem 2, 3, 4, 6, 7
Ecosystem services 216
Environmental Impact Assessment Policy 251
Environmental Liability Policy 250
Environmental Protection Agency (EPA) 219

Exposure 140, 141, 144, 145, 146, 152, 158, 160, 162, 164

F

fatty acids 21
fauna 34
flora 34
food chain 80, 84, 94
food system 111
Full-site soil remediation 168

G

garbage 79
garbage waste 199
geology 108
global economy 81, 102
Global Soil Partnership (GSP) 20
green chemistry 246, 247

H

habitat 2, 4, 6, 13, 14
hardware 168
hazard index 82
Headaches 34
Heavy metals 78, 85, 94
herbicides 78, 79
high-level radioactive waste (HLW) 248
human health 78, 80, 81, 82, 83, 84, 85, 86, 87
hydrocarbons 21, 27, 28, 30, 31
hydrology 108

I

inappropriate waste disposal 108
Industrial Emissions Policy 250
Industrial equipment 169
industrial products 168
Industrial waste 199
inorganic materials 21
insecticides 108, 133, 134, 135, 136, 137
integrated pollution index 82
intensive farming 36, 59, 60, 61, 62
Intergovernmental Technical Panel on Soils (ITPS) 20
International Council of Mining and Metals (ICMM) 170

L

land degradation 108
lead 108, 112, 129, 132, 133, 136
lipids 21
liquid particles 196
liver 35
living organisms 2, 3, 4, 9, 13
low- and middle-income countries (LMICs) 170
low-level radioactive waste (LLW) 248

M

man-made materials 20
Mercury Regulation 251
Metals 21, 27
microbial activity 218, 223
microorganisms 196, 204
Mineral particles 171
molasses 217, 230, 235

N

naphthalene 108
natural resource 2
nausea 34
Nemerow pollution index 82
nitrates 21
Non-polar chemicals 222

Nuclear Safety Commission (NSC) 248

O

Oil Pollution Act (OPA) 248
organic carbon 78
organic matter 143, 144, 196, 197
organisms 2, 4, 6, 7, 9, 12, 13, 14, 17
overgrazing 36

P

packaging 168
Pedons 2
persistent organic pollutants (POPs) 224
pesticides 199, 200, 203, 204, 208
Petroleum hydrocarbons, PAHs 108
phosphates 21
phosphorus 197, 202, 203
physicochemical parameters 223
plain air 217
pollution index 82
pollution load index 82
polycyclic aromatic hydrocarbons (PAH) 224
potassium 197
proteins 21

R

Radioactive waste 248
rainfall 2, 8
Raw material 114
Resource Conservation and Recovery Act (RCRA) 254
root crops 141
rubbish waste 199

S

sediments 196
Sewage Sludge Policy 251
single pollution index 82
social-ecological systems 216
soil contamination 20, 23, 30, 31, 32
Soil degradation 35, 36
soil density 218
soil dust 141, 144, 147
soil particles 140, 141, 143, 144, 153, 154, 155, 156, 158, 162, 163, 164, 165
soil pica 142, 143
soil pollution 20, 21, 22, 23, 29, 30, 31
soil polypedon system 2
Soil protection 246, 250
soil quality 35, 36, 61, 62, 74
soil reaction 2
soil resources 2
solvents 108, 217
Standard equipment 169
substance 20
sulfates 21
sustainability 152

T

terrestrial ecosystem 2
Toxic Substances Control Act (TSCA) 254
toxins 108, 110, 111, 130, 131

U

Urban waste soil pollutants 199

V

vegetable oil 217
vegetables 146, 147, 149, 150, 151, 152, 153, 154
vegetation 2, 7, 11
vomiting 34

W

wastewater management 79
water entering farmlands 149
water management 257
water sources 108, 115
wildfires 36

X

xenobiotic chemicals 108